T0210669

SpringerBriefs in Electrical and Computer Engineering

More information about this series at http://www.springer.com/series/10059

Li Wang • Huan Tang

Device-to-Device Communications in Cellular Networks

 Springer

Li Wang
School of Electronic Engineering
Beijing University of Posts
   and Telecommunications
Beijing, China

Huan Tang
Department of Computer Science
University of California-Davis
Davis, CA, USA

ISSN 2191-8112   ISSN 2191-8120   (electronic)
SpringerBriefs in Electrical and Computer Engineering
ISBN 978-3-319-30679-7   ISBN 978-3-319-30681-0   (eBook)
DOI 10.1007/978-3-319-30681-0

Library of Congress Control Number: 2016934414

Printed on acid-free paper

This Springer imprint is published by Springer Nature
The registered company is Springer International Publishing AG Switzerland

To my family, friends, and students
                     – Li Wang

To my family and friends
                     – Huan Tang

Acknowledgments

The authors sincerely acknowledge the support from the National Natural Science Foundation of China under Grant 61201150 and Grant 61571056, the open research fund of the National Mobile Communications Research Laboratory of Southeast University (China) under Grant 2016D04, and the US National Science Foundation under Grant CIF1321143 and Grant ECCS1307820.

Special thanks to Prof. Sherman Shen who made this book possible. The authors would like to express their deepest gratitude to Prof. Gordon L. Stuber of Georgia Institute of Technology (Gatech); Prof. Zhi Ding and Prof. Bernard C. Levy of the University of California, Davis (UC Davis); and Prof. Zhu Han of the University of Houston (UH) for their advice. The authors would also like to express appreciation to graduate students Huaqing Wu and Ruoguang Li of the Beijing University of Posts and Telecommunications (BUPT), for their strong support during the preparation of this monograph.

Beijing, China   Li Wang
Davis, CA, USA   Huan Tang

Contents

Acronyms

3GPP	Third generation partnership project
AWGN	Additive white Gaussian noise
BS	Base station
BSBL	Block sparse Bayesian learning
CDF	Cumulative distribution function
CDM	Code division multiplexing
CIR	Channel impulse response
CP	Cyclic prefix
CSI	Channel state information
CUE	Cellular user equipment
D2D	Device-to-device
DFT	Discrete Fourier transform
DM-RS	Demodulation reference signal
DUE	Device-to-device user equipment
EPC	Evolved packet core
FDD	Frequency division duplex
GA	Genetic algorithm
GFP	Generalized fractional programming
GS	Gale–Shapley
HK	Hopcroft–Karp
IDFT	Inverse discrete Fourier transform
IPPO	Inverse popularity pairing order
KM	Kuhn–Munkres
LHS	Left-hand side
LPI	Lower probability of intercept
LTE	Long-term evolution
LTE-A	Long-term evolution-advanced
LTE-D	Long-term evolution-direct
MAP	Maximum a posteriori
ML	Maximum likelihood
MRC	Maximal ratio combining

MSE	Mean square error
OFDM	Orthogonal frequency division multiplexing
OFDMA	Orthogonal frequency division multiple access
PAR	Peak-to-average ratio
PDF	Probability density function
PDN	Packet data network
PRACH	Physical random access channel
PRB	Physical resource block
ProSe	Proximity service
PUCCH	Physical uplink control channel
PUSCH	Physical uplink shared channel
QoS	Quality of service
RAN	Radio access network
RB	Resource block
RBG	Resource block group
RTD	Round-trip delay
SCA	Successive convex approximation
SC-FDMA	Single-carrier frequency division multiple access
SINR	Signal-to-interference-plus-noise ratio
SNR	Signal-to-noise ratio
SRS	Sounding reference signal
TDD	Time division duplex
TDM	Time division multiplexing
UE	User equipment
WLAN	Wireless local area network
ZC	Zadoff–Chu

Symbols and Notations

$\boldsymbol{F}_{n_i}$	DFT matrix of size $n_i \times n_i$
$\boldsymbol{y}_{CP}$	Received signal at UE-0
N_{SC}	Total number of SRS combs
$\boldsymbol{z}$	AWGN noise vector
$\boldsymbol{W}_i$	Demapped SRS signal from the SRS comb used by UE i
$\boldsymbol{\Psi}$	Coefficient matrix
$\boldsymbol{g}$	Stacked channel vector
$\mathscr{C}$	Index set of CUEs
$\mathscr{R}$	Resource pool for CUEs
$\mathscr{D}$	Index set of DUEs
$\mathscr{D}^i$	Candidate D2D set for CUE i
$\mathscr{C}^j$	Candidate CUE set for D2D link j
σ^2	Background AWGN power level
μ_{ij}	Indicator variable representing spectrum reusing
x_{ikm}	Fraction of RB chunk k allocated to D2D link i for mode m
p_{ikm}	Transmission power of D2D link i on RB chunk k in mode m
$\boldsymbol{w}_{ikm}$	Normalized transmit precoder of the BS
$\boldsymbol{v}_{ikm}$	Normalized receive precoder of the BS
α	Path loss exponent
$r_{ikm}^{(C)}$	Achievable data rate of CUE i working in mode m when RB chunk k is allocated
$r_{ikm}^{(D)}$	Achievable data rate of D2D link i working in mode m when RB chunk k is allocated
R_i^c	Achievable data rate of uplink CUE i
R_j^d	Achievable data rate of D2D link j
P_k	Transmission power of CUE on RB chunk k
P_e	Transmission power of the BS
P_i^c	Transmission power of uplink CUE i
P_j^d	Transmission power of D2D link j
$P_{i,\max}^c$	Maximum transmission power of uplink CUE i

$P_{j,\max}^d$	Maximum transmission power of D2D link j
ξ_i^c	Achievable SINR of CUE i
ξ_j^d	Achievable SINR of D2D link j
$\xi_{i,\min}^c$	Minimum SINR threshold of CUE i
$\xi_{j,\min}^d$	Minimum SINR threshold of D2D link j
$\Pr_s(i,j)$	Success probability of data transmission for D2D link j sharing resource of CUE i
$v_{\min}^d$	Threshold of success probability of D2D data transmission
$(\cdot)_N$	Modulo-N operation
$(\cdot)^+$	$\max(0,\cdot)$

Chapter 1
Introduction

The proliferation of smart phones in the past decade has contributed to the improvement of productivity and lifestyle quality. Modern-day life now depends heavily on ubiquitous wireless access world with high speed and high quality data transmission. On the other hand, such widely spread use of cellular devices leads to an explosive growth of wireless traffic volume, severely straining the limited cellular bandwidth and capacity. To confront these and related challenges, the third generation partnership project (3GPP) long-term evolution (LTE) continues to standardize technologies with higher data rates, lower latency, and lower power consumption. Presently, LTE-Advanced (LTE-A) supports new technology components for LTE to meet wider range of communication service requirements [1, 2]. Device-to-Device (D2D) communication, as a technology component for LTE-A, allows direct wireless links between mobile users without routing data through a base station (BS) or the core network. Given the rapid growth of data traffic, the shortage of radio spectrum, and the pressing need for battery energy consumption, D2D communications can bring some benefits compared to the conventional infrastructure-based communication. The direct benefits include improved system throughput, increased spectrum efficiency and energy efficiency, and reduced transmission delay, etc.

In this book, we aim to identify and discuss several technical challenges in D2D communications in cellular networks. To start with, we first summarize main D2D standardization activities in 3GPP and identify some significant technical challenges, as illustrated in Chap. 1. Furthermore, Chap. 2 emphasizes some of the important research aspects and technologies to enable D2D communications. Specifically, D2D proximity discovery will be discussed in Chap. 3 since it is a prerequisite for adjacent user devices to communicate via D2D links. Once D2D links are successfully established, mode selection, demonstrated in Chap. 4, needs to be investigated to determine the best communication mode for D2D links to enhance

© The Author(s) 2016
L. Wang, H. Tang, *Device-to-Device Communications in Cellular Networks*,
SpringerBriefs in Electrical and Computer Engineering,
DOI 10.1007/978-3-319-30681-0_1

the system performance. Moreover, mode selection is usually implemented by combining resource management in terms of power control and spectrum allocation to further improve system throughput, as shown in Chap. 5.

1.1 D2D Communications Towards 5G

Generally, 2nd generation (2G) to 4th generation (4G) systems are designed from a network-centric perspective. However, 5G networks (which is expected to be standardized around 2020) do not need to be network-centric and move towards device-centric systems. Naturally, the intelligence at the user equipment (UE) side will be exploited in 5G networks to support D2D connectivity, which is mainly driven by the inherent and strong motivation for operators to offload traffic from the core network. The realization of D2D connectivity represents a real step forward for operators to reduce transmission delay and improve energy efficiency, particularly for networks (e.g., social networking) supporting proximity-based services. In the main body of this book, different design aspects and technical challenges will be discussed for D2D communications towards 5G [3].

1.1.1 History of D2D Standardization Activities

Increasingly, mobile stakeholders such as device manufacturers and network operators are accepting that D2D communications will be a cornerstone of future 5G networks, which drives the standardization of D2D technologies. No later than the 3GPP meeting held in June 2011 [4], the concept of D2D discovery and communication was submitted by Qualcomm. Meanwhile, in the 3GPP meeting held in August 2011, a study item description on LTE Direct (LTE-D) was submitted [5] proposing the study of the service requirement for direct over-the-air LTE D2D discovery and communications. Furthermore, main results on D2D use cases and potential requirements for a network operator controlling discovery and communications between nearby devices were studied in the meeting of November 2011 [6]. Towards the finalization of LTE Release-11, 3GPP initiated the agenda for Release-12, which was initiated at a workshop in June 2012. At that workshop, it was agreed that machine type and short-range communication scenarios can be embraced to accommodate new traffic types [7]. Subsequently, the radio access network (RAN) 58th plenary meeting held in December 2012 agreed to start the study of LTE D2D proximity service (ProSe) [8]. The study on ProSe includes two parts, namely D2D discovery and D2D communication. The study on LTE D2D ProSe mainly focuses on technical details, such as discovery signal design, resource allocation and scheduling, synchronization mechanism, channel models, etc. [9].

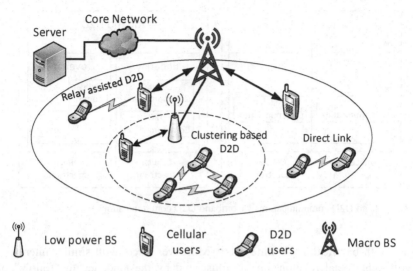

Fig. 1.1 Different use cases of D2D communications

Currently, the overall vision of how the ProSe function is to be implemented has already taken shape, efforts have been made to support infrastructure, such as billing [10]. Most of the architectural progress on ProSe and D2D communications has been summarized in the TS 23.303 document [11]. Today, some of the related ideas on potential license-assisted communication (RP-140770) and on LTE-based short-range radio within licensed bands [12] have already been documented.

1.1.2   Classifications of D2D Communications

Generally speaking, D2D users can communicate with each other in three manners: *D2D direct link*, *relay assisted D2D communications*, and *clustering-based communications*, as shown in Fig. 1.1.

- *D2D Direct Link:* The simplest case of D2D communication occurs when transmitters and receivers exchange data directly with each other without intermediate nodes.
- *Relay Assisted D2D Communications:* Given a scenario where a mobile device wants to connect to another node which is out of its communication coverage or is in a poor channel state with the destination node, cellular users may be employed as relays to improve the data transmission between transmitters and receivers [13].
- *Clustering-Based Communications:* In a content sharing or information diffusion scenario, users requesting the same file in a short range can potentially form a cluster to allow the desired file to be multicasted within the cluster to

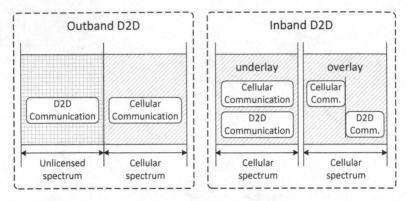

Fig. 1.2 Inband D2D communications vs. outband D2D communications

save both bandwidth and time delay. Moreover, users with similar interests or with tight social relationship (as illustrated by the tendency to "follow" one another) can also form a cluster to share contents with one other via D2D communications [14].

On the other hand, both unlicensed and licensed spectrum resources can be occupied by D2D users for communication, based on which we can further divide D2D communications into *outband D2D communications* and *inband D2D communications* [9]:

- *Outband D2D:* D2D communications under this category exploit unlicensed spectrum [2], as shown in the left side of Fig. 1.2. Key advantages of outband D2D communications lie in the absence of interference between D2D links and cellular links since D2D communications occur on license-exempt bands. Note that, an extra radio interface is necessary for exploiting unlicensed spectrum, which brings up with other wireless technologies together, such as Wi-Fi, Wi-Fi Direct, Bluetooth, and Ultra Wideband technologies [15]. Furthermore, outband D2D can be generally divided into two types, which are *controlled* and *autonomous* communications. In controlled outband D2D communications, coordination between radio interfaces is controlled by cellular networks. In autonomous outband D2D communications, the cellular network controls all the communications but leaves D2D communications to the users themselves. In other words, the second (extra) interface/technology is independent of cellular networks, similar to the current Wi-Fi link. Outband D2D communications face fewer challenges in coordinating the communication resources over two different bands because D2D communications usually happen on a second radio interface. However, only cellular devices with two radio interfaces (e.g., LTE and Wi-Fi) can use outband D2D by simultaneously maintaining D2D communications and cellular communications.

- *Inband D2D:* In this category, D2D communications operate on licensed spectrum (i.e., cellular spectrum) which is also allocated to cellular links [16]. High control over cellular (i.e., licensed) spectrum indicates that it is more convenient to provide better user experiences under a planned environment. Inband D2D communications can be further divided into two types, namely *underlay* D2D communications and *overlay* D2D communications. In underlay D2D communications, DUEs share the same spectrum resources with some other cellular user equipments (CUEs) as shown in the right hand of Fig. 1.2. With spectrum sharing, D2D communications can improve spectrum efficiency and network throughput which are two important performance indices. DUEs working in underlay D2D mode can reuse either CUE uplink or downlink resources (channels) or both in cellular networks. Nevertheless, mutual interference between DUEs and CUEs could critically affect the system performance. On the other hand, D2D links working in overlay communication mode are allocated dedicated cellular resources. At the expense of lower spectral efficiency, overlay D2D communications can usually provide better system performance without co-channel interference under dedicated resources. However, overlay D2D communications are not as efficient as underlay in terms of spectrum efficiency.

The rest of our book focuses on inband D2D communications, thanks largely to its potential to guarantee QoS, by allocating power and spectrum to control interference.

1.1.3 D2D-Assisted Cellular Communication

Recall that D2D communications using cellular spectrum under the control of cellular infrastructure enable devices to communicate directly without intermediate nodes. As mobile applications exploiting the proximity of mobile devices continue to gain popularity, D2D communications in cellular networks also attract more and more attention owing to their ability to share data at higher speed via local links. The emergence of D2D communications in cellular networks promises multiple performance benefits [17]. Generally, D2D can support communications at higher bit rates, lower delays, while consuming less energy. For example, two adjacent devices, DUE 1 and DUE 2 in Fig. 1.3, can usually achieve better performances when communicating directly by setting up a D2D link rather than seeking relay help from BS 1. In particular, it is more resource-efficient for adjacent DUEs to communicate directly with one other as opposed to routing data through a BS or even the core network. In other words, compared to the traditional cellular communications which occupy both uplink and downlink resource, D2D communications save energy and improve spectrum efficiency. Furthermore, switching from an infrastructure path to a direct path may offload the cellular and backhaul traffic, alleviate network congestion, thereby benefiting other CUEs in the network; see,

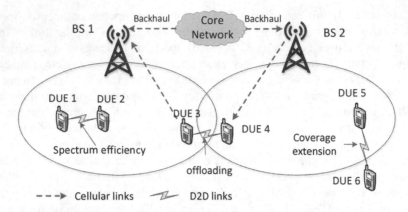

Fig. 1.3 Potential benefits of D2D communications

e.g., the D2D communication link between DUE 3 and DUE 4 as shown in Fig. 1.3. Other benefits include coverage extension. For example, DUE 6 can accomplish data transmission by communicating with DUE 5 via D2D links even though it is out of the coverage of BSs.

1.2 Research Challenges in D2D-Assisted Networks

To facilitate local area services, D2D communications present a promising technology in future wireless networks to improve network capacity and user experience. In 3GPP Release 12, it is clear that D2D technology is of high interest for future development investigation. As explained earlier, mobile network operators can derive substantial benefit by leveraging the coordinated and network-assisted D2D technologies. However, introducing D2D technology into today's network infrastructure engenders a number of challenges and requires functional updates to the current longstanding cellular architecture, i.e., synchronization, device discovery, mode selection, interference management, power control, and channel measurements. In the following part of this section, we introduce some research challenges in D2D communications.

1.2.1 Synchronization

In an LTE network-assisted D2D scenario, two user equipments (UEs) of a D2D pair are synchronized with the BS, implying that slot and frame timing as well as frequency synchronization are acquired and aligned [3]. Synchronized D2D transmissions are appealing since BSs can provide synchronization beacons.

For instance, continuous searching for discovery signals is required in asynchronous discovery schemes. However, UEs can be active only during predetermined time slots for receiving discovery related signals in the process of time synchronized device discovery, which consumes significantly less energy [18].

However, it is challenging to realize synchronization between two UEs because of at least two reasons [18]:

- Two UEs that form a D2D link pair may be associated with different BSs that are not synchronized;
- Even UEs are located in the same cell, the distance between different UEs to the BS may be different, and the application of different timing advance adjustments may be required.

Therefore, further study is required to investigate the impact of timing misalignment on system performance, and additional synchronization methods are needed if the impact turns out to be non-negligible.

1.2.2 Device Discovery

Before communicating, UEs have to search for potential devices in proximity for D2D communication and determine the identification of the discovered peers. Therefore, device discovery is a key building block in D2D-based networks. Proximity device discovery implies the ability for a UE to continuously search for relevant peers within its physical proximity. To launch a proximity discovery procedure, only valid D2D links (with successful proximity discovery procedure) are considered in the subsequent D2D communication procedures.

A crucial question in device discovery is how to discover suitable peers to establish D2D links in a fast and efficient manner, so as to achieve an optimal tradeoff or to balance the tradeoff among system throughput, interference control (elimination), and resource utilization efficiency.

- *Discovery Signal (Pilot) Design:* In the procedure of device discovery, discovery signals transmitted by a UE may be detected by other UEs. Obviously, information carried by the discovery signals should be scheduled carefully. For example, the amount of information sent by UEs during discovery procedure may impact many design factors, e.g., the required amount of radio resources and the design of discovery signal or channel structure. Thus, a rough estimate of the quantity of discovery information is required to facilitate the design [18].
- *Synchronous vs. Asynchronous Discovery:* Generally, synchronous discovery is more appealing when compared to asynchronous schemes because of its potential to achieve higher spectrum efficiency and lower energy consumption. Moreover, synchronous discovery schemes can usually result in more reliable and faster discovery. However, it is questionable to assume synchronization a priori before device discovery for those UEs that are out of the coverage of cellular networks [18].

1.2.3   Mode Selection

In traditional cellular networks, UEs communicate with others through the BS by occupying both uplink and downlink channels. When D2D communications are made possible, UEs can decide among multiple mode choices for communications, which we refer to as mode selection. In mode selection, the best mode can be selected to optimize the system performance by taking the network load, channel condition, and interference situation into consideration. Challenges exist in mode selection since the design of the schemes has to consider [9]:

- How often we should implement mode selection and associated channel quality estimations. When considering the rapidly evolving radio conditions within the cell and between the D2D pairs, timescale for mode selection cannot be too coarse. On the other hand, the measurements and control signaling required for mode selection should be kept at a minimum to avoid too much overhead [15].
- What measurements, reporting mechanisms, and (periodic and/or event triggered, hybrid) algorithms the devices and BSs should use to select different communication modes [15].

1.2.4   Interference Management

When working in cellular mode or dedicated D2D mode, users are allocated with orthogonal spectrum, which can lead to inefficient use of available spectrum resources. To improve spectrum efficiency, D2D links can communicate by reusing the same spectral resources known as physical resource blocks (PRBs) with cellular links. However, co-channel interference brought by spectrum sharing must be coordinated carefully to guarantee the required QoS for those UEs involved. In such a co-channel sharing mode, interference scenarios are different when D2D communications may reuse cellular downlink or uplink channel resources [19].

When downlink resources are reused by D2D links, D2D receivers, and CUEs using the same spectrum resources would interfere. For DUEs reusing the downlink spectrum, interference comes from other co-channel DUEs and the BS. Since a D2D pair is usually formed between UEs in proximity, the power needed for D2D communications is much lower than that for traditional cellular communications. Therefore, DUEs in this case have to stay far away from high-power BS to prevent themselves from being the victims of overwhelming interference power. For CUEs providing downlink resources, the interference comes from all other co-channel DUEs. Thus, DUEs have to keep a distance from the CUEs to avoid causing harmful interference to the regularly connected CUE downlinks.

On the contrary, BS and D2D receivers are also interference victims when DUEs share the uplink cellular resources. In this case, DUEs have to keep themselves off the co-channel CUEs to avoid suffering debilitating interferences. For CUEs on the uplink, the stationary BS is the receiving node which must coordinate the DUEs and control the interference from all the co-channel DUEs.

In addition to intra-cell interference, intercell interference coordination has also become a major design question in cellular networks supporting D2D communications. Interference management in this case is rather complex since interference needs to be coordinated between multiple cells and between cellular and D2D layers also.

1.2.5   Power Control

In D2D-assisted networks, power control is useful for interference mitigation, energy conservation, and throughput maximization, etc. [20]. Generally, one critical problem of power control design in network-assisted D2D communications is how to coordinate the involved BSs and devices, as well as the timescale of interaction between the network and DUEs [9]. One approach is to let the BS schedule dynamically on a very small timescale, whereas an alternative design is to let the BS take charge of long-term power control and allow D2D pairs to schedule their transmit powers autonomously. Allowing DUEs themselves concern control capabilities with respect to their own transmit power, control signaling overhead and delay may be reduced. For example, the BS may just be responsible for open loop power control, and setting a rough transmit power level and permissible power range, while DUEs can handle finer closed loop power control in accordance to rapid mobile channel variations [18].

1.2.6   Channel Measurements

To implement power control and interference mitigation, channel measurement is required to inform networks channel conditions of involved users. To enable channel measurement, the design of reference signals used for D2D links requires further study, although initial channel measurement may be performed during the process of device discovery by exploiting discovery signals. Generally, measurement of the received strength of reference signals transmitted by the BS on the downlink can be used to estimate the interference caused by D2D communications. Thus, these measurement reports can be useful for the BS to assign resources for D2D links. In the uplink, two types of reference signals are usually used: sounding and demodulation reference signals [3].

The sounding reference signals (SRS) are usually transmitted on a wider bandwidth than the actual data transmission to facilitate estimating channel information. This may also be useful in D2D-assisted networks which have strict control of resource allocation. On the other hand, the demodulation reference symbols (DMRS), which are transported beside the PRBs of the payload, can be used for demodulation, channel estimation, and channel equalization, and could possibly serve the same purpose in D2D communications [3]. However, it remains to be open whether other reference signals may be better suited for D2D communications.

1.3 Outline of the Book

This book is organized into the following structure. Chapter 1 presents an overview tutorial on D2D communications, including the essentiality and potential advantages of D2D communications, as well as design challenges on D2D communications. After the introduction, some design aspects and key technologies in D2D communications are illustrated in Chap. 2, including D2D proximity discovery, mode selection, and resource management. D2D proximity discovery, as elaborated in Chap. 3, is a prerequisite for the subsequent D2D communication. Specifically, we focus on neighbor UE discovery utilizing the SRS channel, which can be accessed by peer UEs that are LTE-compliant. After proximity discovery, mode selection is executed for potential D2D links to determine the best communication mode to enhance the system performance. As illustrated in Chap. 4, mixed-mode D2D communications are investigated to maximize the D2D sum rate under cellular rate constraints, in which D2D links can operate in multiple modes through resource multiplexing. The process of mode selection involves both D2D overlay case, where D2D links are given dedicated spectrum resources, and D2D underlay case, where DUEs share the same spectrum resources with CUEs. To provide better assistance for mode selection, resource management in terms of power control and spectrum allocation can be implemented to further improve system throughput. Specifically, Chap. 5 studies the problem of resource management in D2D underlaid cellular network to maximize the system data rate by sharing uplink spectrum resources. Under different user requirements, different bipartite graphs can be formed to solve different resource allocation problems. Moreover, noting the fact that D2D links are not always stable because of user mobility, resource management can be executed by jointly considering social characteristics of user nodes and physical channel conditions. Finally, Chap. 6 provides the summary of the technical contents discussed in this book.

References

1. K. Doppler, M. Rinne, C. Wijting, C.B. Ribeiro, K. Hugl, Device-to-device communication as an underlay to LTE-advanced networks. IEEE Commun. Mag. **47**(12), 42–49 (2009)
2. A. Asadi, Q. Wang, V. Mancuso, A survey on device-to-device communication in cellular networks. IEEE Commun. Surv. Tutorials **16**(4), 1801–1819 (2014)
3. S. Mumtaz, K.M.S. Huq, J. Rodriguez, Direct mobile-to-mobile communication: paradigm for 5G. IEEE Wirel. Commun. **21**(5), 14–23 (2014)
4. Qualcomm Incorporated, On the need for a 3GPP study on LTE Device-to-Device discovery and communication, 3GPP SP-110383, June 2011
5. 3GPP, Study on LTE direct, 3GPP S1-112017, August 2011
6. 3GPP, Study on Proximity-based services, 3GPP SP-110590, September 2011
7. 3GPP, 3rd generation partnership project; technical specification group SA; feasibility study for proximity services (ProSe) (Release 12), TR 22.803 V1.0.0, August 2012
8. Qualcomm, Study on LTE device to device proximity services, 3GPP RP-122009, December 2012

9. S. Mumtaz, *Smart Device to Smart Device Communication*, ed. by J. Rodriguez (Springer, New York, 2014)
10. 3GPP TS 32.277, Telecommunication management; charging management; proximity-based services (ProSe) charging, August 2014
11. 3GPP TS 23.303, Proximity-based services (ProSe); stage 2, February 2014
12. 3GPP TR 36.843, Study on LTE device to device proximity services; radio aspects, March 2014
13. L. Lei, Z. Zhong, C. Lin, X. Shen, Operator controlled device-to-device communications in LTE-advanced networks. IEEE Wirel. Commun. **19**(3), 96–104 (2012)
14. C. Cao, L. Wang, M. Song, Y. Zhang, Admission policy based clustering scheme for D2D underlay communications, in *2014 IEEE 25th International Symposium on Personal, Indoor and Mobile Radio Communications (PIMRC)*, Washington, DC, September 2014, pp. 1937–1942
15. G. Fodor, E. Dahlman, G. Mildh, S. Parkvall, N. Reider, G. Miklós, Z. Turányi, Design aspects of network assisted device-to-device communications. IEEE Commun. Mag. **50**(3), 170–177 (2012)
16. L. Lei, Y. Kuang, X. Shen, C. Lin, Z. Zhong, Resource control in network assisted device-to-device communications: solutions and challenges. IEEE Commun. Mag. **52**(6), 108–117 (2014)
17. K. Doppler, M.P. Rinne, P. Jänis, C. Ribeiro, K. Hugl, Device-to-device communications; functional prospects for LTE-advanced networks, in *Proceedings of IEEE International Conference on Communications Workshops. ICC Workshops 2009*, June 2009, pp. 1–6
18. X. Lin, J.G. Andrews, A. Ghosh, R. Ratasuk, An overview of 3GPP device-to-device proximity services. IEEE Commun. Mag. **52**(4), 40–48 (2014)
19. L. Wei, R.Q. Hu, T. Qian, G. Wu, Enable device-to-device communications underlaying cellular networks: challenges and research aspects. IEEE Commun. Mag. **52**(6), 90–96 (2014)
20. G. Fodor, D.D. Penda, M. Belleschi, M. Johansson, A. Abrardo, A comparative study of power control approaches for device-to-device communications, in *Proceedings of 2013 IEEE International Conference on Communications*, June 2013, pp. 9–13

Chapter 2
Critical Technologies for D2D Communications

D2D communications is a relatively new study item, and its relevant critical consideration and technologies are largely under development, as discussed in Sect. 1.2. Exploiting D2D communications in cellular networks promotes new promising ways to operate networks in a highly efficient manner. In the following, we describe some key points required to enable D2D communications, and categorize them under three topics: proximity discovery, mode selection, and resource management.

2.1 Proximity Discovery

Recall that the discovery of proximity devices is a prerequisite for initiating the D2D communications. During the discovery phase, UEs try to discover potential candidates and prepare to establish direct communication links.

2.1.1 Taxonomy of D2D Proximity Discovery

Generally, existing proximity discovery approaches for D2D communications can be classified into two types:

Distributed Approaches In this category, UEs broadcast their identities periodically so that other UEs may be aware of their existence and decide whether to start D2D communications or not. This approach is distributed and does not need the involvement of the BS. Distributed proximity discovery approaches are flexible, but time and energy consuming, employing beacon signals and sophisticated scanning.

© The Author(s) 2016
L. Wang, H. Tang, *Device-to-Device Communications in Cellular Networks*,
SpringerBriefs in Electrical and Computer Engineering,
DOI 10.1007/978-3-319-30681-0_2

Network-Assisted Approaches In network-assisted proximity discovery, pairable D2D UEs detect and identify each other with the assistance of the network. One UE informs the BS about its intension to communicate with another partner UE and transmits its own beacon signal. Then the BS executes message exchanges to acquire the identity and information about the potential link [1]. This approach can be either *centralized* or *semi-centralized*, depending on the involvement of BS during the procedure of proximity discovery [2].

2.1.2   Procedure of Proximity Discovery

Despite the fact that UEs are able to detect proximate devices autonomously, flexibly, and scalably, network-assisted discovery guarantees more predictable performance because the operator will be responsible for resource control and interference coordination. Compared with the distributed approaches, several important advantages exist in network-assisted proximity discovery schemes. As mentioned before, peer discovery without network support is typically time-and-energy-consuming, by using beacon signals and sophisticated scanning, especially in a high load system. Furthermore, unexpected interference can be reduced if the discovery procedure is coordinated by the network. With these in mind, discovery procedures with network assistance are considered in this book.

In centralized network-assisted proximity discovery, where BS coordinates all the messages in every step, UEs transmit or listen only upon request from the base station [3]. As shown in Fig. 2.1, the procedure of centralized network-assisted proximity discovery can be described as:

Step 1:   UE 1 informs the BS about its intention to communicate with UE 2;
Step 2:   BS requests UE 2 to expect a discovery message from UE 1, and UE 1 is
          informed to send the discovery message;
Step 3:   UE 1 sends discovery message to UE 2;
Step 4:   UE 2 reports the measured signal-to-interference-plus-noise ratio (SINR)
          value of the message transmitted by UE 1 to the BS;
Step 5:   BS requires both UE 1 and UE 2 to listen for interference from existing
          users in the cell;
Step 6:   UE 1 and UE 2 report to the BS their measured results of interference;
Step 7:   BS instructs UE 1 and UE 2 to communicate with each other via D2D link
          if the direct link between UE 1 and UE 2 is favorable.

In semi-centralized proximity discovery, the role of the BS is less dominant since the initial steps of the procedure do not include message transmissions with the BS. As shown in Fig. 2.2, the procedure of semi-centralized proximity discovery can be described as:

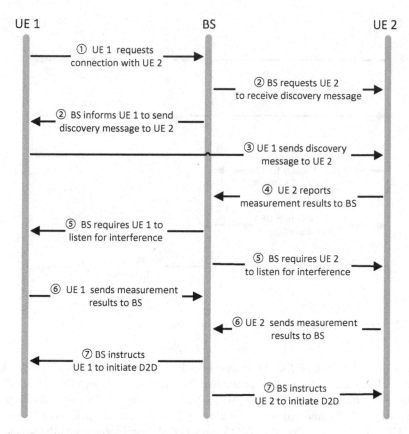

Fig. 2.1 Centralized proximity discovery procedure

Step 1:   UE 1 sends discovery message to UE 2 without requesting permission
          from the BS. Both UE 1 and UE 2 listen for interference from the other
          users in the cell and estimate their path gains to the BS immediately upon
          the reception of the message sent by UE 1 to UE 2;
Step 2:   UE 2 reports its measured results to UE 1 in terms of the interference it
          suffered and its path gains to the BS;
Step 3:   UE 1 feeds back to the BS about the SINR and interference measurements
          for both UE 1 and UE 2;
Step 4:   BS requests both UEs to initiate D2D communications.

2.1.3   Related Works and Motivations

To design proximity discovery algorithms, four requirements should be considered,
i.e., high energy efficiency, long discovery range, low channel resource cost, and
minimum interference caused by the transmission of discovery beacons. Generally,

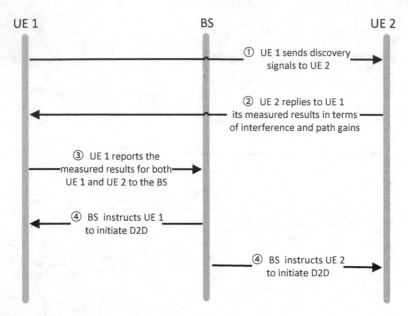

Fig. 2.2 Semi-centralized proximity discovery procedure

there exist location-based D2D discovery schemes and beacon-based D2D discovery schemes. Location-based proximity discovery exploits wireless localization
methods [4], e.g., angle-of-arrival (AOA), time-of-arrival (TOA), time-difference-
of-arrival (TDOA), and global positioning system (GPS), to track the location of
each UE [5]. However, the AOA, TOA, and TDOA cannot guarantee the accuracy
of detection, and the GPS-based localization can neither be implemented by the
UEs without GPS devices nor by the UEs where the GPS signal cannot reach. The
beacon-based D2D proximity discovery, however, enables UEs to send beacons to
nearby devices. In [6], D2D discovery in wireless local area network (WLAN) has
been investigated in which a DUE can periodically send a beacon message to be
discovered by other devices. The authors of [7] have proposed a D2D discovery
protocol for FlashLinQ, where each device transmits a discovery signal. However,
the aforementioned peer discovery is performed without network support, which
costs considerable time and energy.

In network-assisted D2D discovery, networks can recognize whether two devices
can establish a D2D link and mediate the discovery process. Moreover, networks
can coordinate time, frequency, and coding scheme for devices to send beacons [8].
Authors of [8] have categorized the network-assisted discovery into a priori
discovery and a posteriori discovery according to the timing relationship between
discovery phase and communication phase. In network-assisted discovery, it is
a design goal to make such peer discovery and pairing procedures faster, more
efficient in terms of energy consumption and more user friendly [9]. However, additional resources are required to implement D2D neighbor discovery as discussed in
[7, 10]. Therefore, it is necessary to propose novel schemes operating on existing
channels to facilitate proximity discovery.

2.2 Mode Selection

Recall that D2D communications have the potential to achieve performance gain over traditional cellular communications in terms of bandwidth, time delay, and power. Nevertheless, it does not mean that it is always optimal for UEs to work in D2D mode from the performance perspective. In D2D-assisted cellular networks, two UEs can operate in regular cellular mode communicating through the BS or in D2D mode communicating directly with each other. Furthermore, when working in D2D mode, data transmission can be achieved by sharing the spectrum of CUEs or by using a dedicated cellular spectrum. Those UEs can switch between two modes by considering QoS requirements. Generally, two UEs requesting for communication can work on one mode or hybrid modes based on practical requirements.

2.2.1 Taxonomy of Typical Communication Modes

Basically, UEs can work in one of the four modes as illustrated in Fig. 2.3.

D2D Silent Mode When available resources are not enough for D2D communications with dedicated resources, and spectrum reuse is impossible either owing to harmful interference, D2D users are incapable of data transmission and have to keep silent.

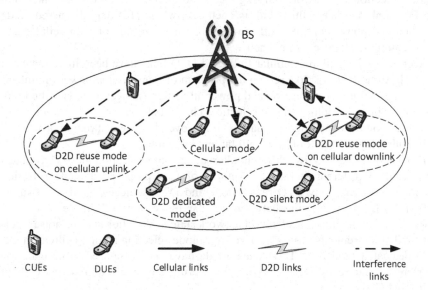

Fig. 2.3 D2D communication modes

D2D Reuse Mode UEs communicate directly via D2D links by sharing the uplink or downlink spectrum resources of CUEs in cellular D2D underlay.

D2D Dedicated Mode Dedicated cellular spectrum resources are allocated for UEs to communicate directly via D2D links.

Cellular Mode In this mode, two UEs can communicate with each other through the BS without co-channel spectrum sharing in traditional way.

Practically, different communication modes can be selected according to channel condition variation and service requirements. D2D reuse mode performs better in terms of spectrum efficiency, but the co-channel interference caused by spectrum sharing is challenging. On the other hand, D2D dedicated mode and cellular mode can be chosen to ease the task of interference management and achieve better user experience with less satisfying spectrum efficiency. By selecting suitable transmission modes for potential D2D links, the overall network performance can be optimized while guaranteeing QoS requirements of involved users.

2.2.2 Related Works and Motivations

There are already a number of mode selection schemes including those in [11–16], by considering channel conditions, power constraints, and system QoS requirements. Specifically, a simple scenario where only one D2D link and one cellular link exist in each single cell has been investigated in [11, 12]. Authors of [11] considered a mode selection scheme for both single-cell and multiple-cell scenarios, including one D2D link and one cellular link in each cell, where [12] only discussed mode selection and power control jointly to optimize the sum rate of both cellular and D2D communications in a single cell.

Scenarios with multiple cellular links and D2D links have been investigated in [13, 14]. Authors of [13] focused on the problem of transmission power minimization through joint mode selection and power allocation, subjecting to specified link rate constraints. In addition, mode selection, channel allocation, and power control have been jointly considered to maximize the overall system throughput in [14], while satisfying the SINR of both cellular and D2D links.

From another perspective, different alternative communication modes have been considered in different literatures. Authors of [15] proposed a mode selection scheme considering only D2D reuse mode and D2D dedicated mode in a single cell. In [16], mode selection between cellular mode and D2D reuse mode was proposed by taking into consideration network state information such as noise levels and SINR requirements of users. Moreover, mode selection among cellular mode, D2D dedicated mode, and D2D reuse mode have been proposed to mitigate the interference caused by D2D links by considering the QoS requirements of both DUEs and CUEs [11, 14].

However, existing works usually focus on *binary mode selection*, where each D2D link can only operate in one mode by using a binary mode indicator to indicate the selection of a certain mode. Design of *mixed-mode operation* is of practical interest where each D2D link can utilize multiple modes through resource multiplexing. Practically, when both D2D and cellular links must meet certain QoS objectives, mixed-mode operation may be preferable since it can leverage advantages of different modes. Therefore, researches on mixed-mode selection strategies would be meaningful to satisfy differentiated cellular and D2D link requirements.

2.3 Resource Management

During mode selection, resource management is usually implemented simultaneously to evaluate achievable performance when working on different modes. Regarded as one of the most critical issues in wireless networks, resource management can be carried out to facilitate interference mitigation, energy conservation, and throughput maximization, among others.

2.3.1 Related Technologies in Resource Management

Generally, resource management in network-assisted D2D communications mainly consists of two parts: spectrum allocation and power control, which are elaborated as follows:

Spectrum Allocation In 3GPP LTE specifications, UEs are allocated with certain number of subcarriers for a predetermined amount of time duration, which are defined as physical resource blocks or PRBs. Each PRB occupies one slot in time domain and 180 kHz in frequency domain, i.e., 12 subcarriers with subcarrier spacing of 15 kHz. A PRB is the smallest unit of spectral resource that the BS can allocate. The allocation of cellular resources to the D2D communications is critical since the interference to other CUEs and DUEs should be kept below a certain level to guarantee QoS requirements. Usually, resource allocation is jointly considered with mode selection, to determine whether some dedicated PRBs or shared PRBs should be allocated to D2D communications. When DUEs communicate in underlay mode, it should be determined which CUE's resource should be shared, on the other hand, how many PRBs permitted for D2D communications should be considered if they work in an overlay mode [17].

Power Control Power control is a key mechanism to mitigate interference among users in D2D-assisted networks. Basically, transmit power of involved users should be controlled to guarantee QoS requirements (e.g., SINR requirements) of different users in the network. Power control can be carried out to maximize the overall network throughput, which means that in some cases, we may need to lower the transmit power of cellular links to improve the achievable data rate of DUEs and further enhance the overall sum rate in the network. Meanwhile, the CUE performance should also be guaranteed. In addition, given the fact that mobile devices rely on their limited battery energy for their operation, power conservation is also vital in D2D communications [17]. Therefore, power control is usually implemented considering the tradeoff between energy saving and achievable throughput.

Generally, spectrum allocation and power control are inseparable and intertwined. When shared spectrum resources are allocated, mutual interference exists between the co-channel 2CUEs and DUEs using the same resource in cellular D2D underlay. In this case, transmit power of DUEs should be reduced according to the backoff value from the transmit power determined by cellular power control [18]. In addition, BS can set the maximum D2D transmit power to a predetermined value, which can be set according to long-term observations for the impact of different D2D power levels on the quality of links. When assigned dedicated spectrum resources, however, no backoff value is required and the maximum transmit power for DUEs can be usually higher than that in spectrum sharing case.

Basically, resource management in D2D-assisted cellular networks can be implemented in a *centralized* manner or a *distributed* manner. In centralized resource management, BS assumes the full responsibility in controlling the resources of D2D communications. On the contrary, DUEs sense the network environment and adaptively utilize the resources without causing too much interference to other users in distributed manner. Centralized resource management can usually achieve more predictable and better performance, but resulting in high complexity. In contrast, less optimal solutions may be found in distributed manner with higher flexibility, lower complexity, and improved scalability.

2.3.2   Related Works and Motivations

There have been substantial researches on resource management in D2D-assisted cellular networks. Resource management in cellular D2D overlay has been studied in [8, 19, 20]. Specifically, BS-assisted scheduling and D2D power control scheme have been proposed in [8] to increase energy efficiency, and [19] developed a spectrum sharing protocol targeting a cellular D2D overlay network to improve the system sum rate.

Different from the overlay cases, resource management in underlaying D2D case concerns more about mutual interference caused by spectrum sharing. There have been a variety of metrics used in the literature to evaluate the effect of resource management, such as spectrum efficiency, power efficiency, maximization of allowable D2D links, fairness improvement, etc. The improvement of spectrum efficiency has been investigated in [21, 22]. Specifically, Xu et al. [21] proposed a new interference cancellation scheme based on the location of users, and a dedicated control channel has been allocated for D2D users to avoid using resource blocks which interfere with CUEs. The work in [22] investigated interference management by defining an interference limited area where no CUEs can occupy the same resources as D2D pairs. Moreover, power efficiency enhancement by proper resource management has been studied in [13, 23]. A heuristic algorithm proposed in [13] focuses on power allocation and mode selection in orthogonal frequency division multiple access (OFDMA) based cellular networks. Authors of [23] studied means to minimize the overall transmission power in a multi-cell orthogonal frequency division multiplexing (OFDM) cellular network by formulating a problem of joint mode selection, spectrum allocation, and power allocation through linear programming. In addition, the maximization of the number of admitted D2D links has been considered in a single-cell network in [24]. Fairness index is another metric considered in resource management for D2D communications [25]. Similarly, resource management for D2D underlaying communications focusing on other metrics has also been investigated, such as cellular coverage extension, reliability improvement, and traffic offloading.

However, most of the existing works on resource management have an ideal assumption that D2D links are always stable. However, D2D links may experience intermittent outage in practice due to user mobility. Exploiting social relationship and interaction among mobile users can facilitate the formation of stable D2D links [26, 27]. Therefore, resource management schemes considering both physical channel condition and social behavior of users should be investigated to improve system reliability and robustness.

Practically, resource management in D2D-assisted cellular networks often involves integer programming problems, such as partner selection and channel allocation. For instance, a DUE can choose a partner to form a D2D link considering different requirements, e.g., whether the chosen partner has the target content it needs, whether the physical link is stable enough for data transmission, and whether the selected partner can be fully trusted or tightly connected. In addition, spectrum resource allocation for each D2D link can be implemented to meet certain objectives, e.g., to ensure QoS requirements for both D2D links and CUEs, to maximize potential number of D2D links in cellular D2D underlay networks, to guarantee that the pairing between D2D links and CUEs' resources is stable and robust in the long term, and to optimize the overall sum data rate or secrecy rate for all the D2D links and CUEs in the system. All these demands can be satisfied by selecting appropriate partners, which can be formulated as a *bipartite graph* problem [28]. Well studied in graph theory, bipartite graph matching provides efficient solutions to resource allocation problems based on each entity's requirements.

Applying graph theory to resource allocation in a network is not a new concept. However, this approach is scattered in the literature such as in cognitive networks and relay-assisted cooperative networks [29, 30]. One of very early connections between bipartite graph and resource management in networks has been reported in [31], which links network control algorithms and edge coloring algorithms for bipartite graphs. As a matter of fact, bipartite graph matching has been thoroughly studied and widely applied to discrete resource allocation in social economics for a long time [32, 33], while its broad applications include house assignment, hospital bed matching, and college admission/selection, among others. A bipartite graphical method has been introduced to dynamic resource allocation in wireless mesh networks to simultaneously consider both bandwidth utilization and starvation problems [34]. The idea of leveraging bipartite graphs in resource management can help the resource management in D2D-assisted networks. Utilizing graph theory, one such work focused on pairing D2D links and CUEs to maximize system overall throughput [12, 35]. Therefore, additional problems in resource management in D2D-assisted cellular networks leveraging bipartite graph are also topics worth studying.

2.4 Chapter Summary

Network-assisted D2D communications in cellular spectrum can take advantage of the proximity of communication devices, allow resource sharing between D2D pairs and cellular users, and reap further rewards such as multi-hop network gain. To harvest these potential gains, there is a need to carefully design inter-connected mechanisms. In the remaining chapters of this book, design aspects addressed in this chapter including proximity peer discovery, mode selection, and resource management will be addressed in greater detail.

References

1. K. Doppler, M. Rinne, C. Wijting, C.B. Ribeiro, K. Hugl, Device-to-device communication as an underlay to LTE-advanced networks. IEEE Commun. Mag. **47**(12), 42–49 (2009)
2. S. Mumtaz, in *Smart Device to Smart Device Communication*, ed. by J. Rodriguez (Springer, New York, 2014)
3. A. Thanos, S. Shalmashi, G. Miao, Network-assisted discovery for device-to-device communications, in *Proceedings of IEEE GLOBECOM Workshops*, Atlanta, GA, December 2013, pp. 660–664
4. S. Gezici, A survey on wireless position estimation. Wirel. Pers. Commun. **44**(3), 263–282 (2007)
5. K.W. Choi, Z. Han, Device-to-device discovery for proximity-based service in LTE-advanced system. IEEE J. Sel. Areas Commun. **33**(1), 55–66 (2015)
6. S. Lu, S. Shere, Y. Liu, Y. Liu, Device discovery and connection establishment approach using ad-hoc Wi-Fi for opportunistic networks, in *Proceedings of IEEE PERCOM*, Seattle, WA, March 2011, pp. 461–466
7. F. Baccelli, N. Khude, R. Laroia, J. Li, T. Richardson, S. Shakkottai, S. Tavildar, X. Wu, On the design of device-to-device autonomous discovery, in *Proceedings of Fourth International Conference on Communication Systems and Networks*, January 2012, pp. 1–9
8. G. Fodor, E. Dahlman, G. Mildh, S. Parkvall, N. Reider, G. Miklós, Z. Turányi, Design aspects of network assisted device-to-device communications. IEEE Commun. Mag. **50**(3), 170–177 (2012)
9. E. Dahlman, G. Fodor, G. Mikls, in Design aspects of network assisted D2D communications in LTE. Technical report in Ericsson. EAB-11: 035391 (2011)
10. B. Kaufman, B. Aazhang, J. Lilleberg, Interference aware link discovery for device to device communication, in *Proceedings of Asilomar Conference on Signals, Systems and Computers*, November 2009

11. K. Doppler, C.H. Yu, C.B. Ribeiro, P. Jänis, Mode selection for device-to-device communication underlaying an LTE-advanced network, in *Proceedings of IEEE Wireless Communications and Networking Conference (WCNC)* (2010), pp. 1–6,
12. C. Yu, K. Doppler, C.B. Ribeiro, O. Tirkkonen, Resource sharing optimization for device-to-device communication underlaying cellular networks. IEEE Trans. Wirel. Commun. **10**(8), 2752–2763 (2011)
13. X. Xiao, X. Tao, J. Lu, A QoS-aware power optimization scheme in OFDMA systems with integrated device-to-device (D2D) communications, in *Proceedings of IEEE Vehicular Technology Conference (VTC Fall)*, September 2011, pp. 5–8
14. G. Yu, L. Xu, D. Feng, R. Yin, G.Y. Li, Y. Jiang, Joint mode selection and resource allocation for device-to-device communications. IEEE Trans. Commun. **62**(11), 3814–3824 (2014)
15. Z. Liu, T. Peng, S. Xiang, W. Wang, Mode selection for device-to-device (D2D) communication under LTE-advanced networks, in *Proceedings of IEEE International Conference on Communications (ICC)* (2012), pp. 5563–5567
16. S. Hakola, T. Chen, J. Lehtomaki, T. Koskela, Device-to-device (D2D) communication in cellular network-performance analysis of optimum and practical communication mode selection, in *Proceedings of IEEE WCNC*, Sydney, April 2010, pp. 1–6
17. L. Wei, R.Q. Hu, T. Qian, G. Wu, Enable device-to-device communications underlaying cellular networks: challenges and research aspects. IEEE Commun. Mag. **52**(6), 90–96 (2014)
18. P. Jänis, C-H. Yu, K. Doppler, C. Ribeiro, C. Wijting, K. Hugl, O. Tirkkonen, V. Koivunen, Device-to-device communication underlaying cellular communications systems. Int. J. Commun. Netw. Syst. Sci. **2**(3), 169–78 (2009)
19. Y. Pei, Y.-C. Liang, Resource allocation for device-to-device communications overlaying two-way cellular networks. IEEE Trans. Wirel. Commun. **12**(7), 3611–3621 (2013)
20. L. Wang, H. Wu, Jamming partner selection for maximising the worst D2D secrecy rate based on social trust. Trans. Emerg. Telecommun. Technol. (2015). doi:10.1002/ett.2992
21. S. Xu, H. Wang, T. Chen, Q. Huang, T. Peng, Effective interference cancellation scheme for device-to-device communication underlaying cellular networks, in *Proceedings of IEEE Vehicular Technology Conference Fall* (2010), pp. 1–5
22. H. Min, J. Lee, S. Park, D. Hong, Capacity enhancement using an interference limited area for device-to-device uplink underlaying cellular networks. IEEE Trans. Wirel. Commun. **10**(12), 3995–4000 (2011)
23. M. Belleschi, G. Fodor, A. Abrardo, Performance analysis of a distributed resource allocation scheme for D2D communications, in *Proceedings of IEEE GLOBECOM Workshops* (2011), pp. 358–362
24. T. Han, R. Yin, Y. Xu, G. Yu, Uplink channel reusing selection optimization for device-to-device communication underlaying cellular networks, in *Proceedings of IEEE PIMRC* (2012), pp. 559–564
25. C. Xu, L. Song, Z. Han, Q. Zhao, X. Wang, B. Jiao, Interference-aware resource allocation for device-to-device communications as an underlay using sequential second price auction, in *Proceedings of IEEE ICC*, Ottawa, ON, June 2012, pp. 445–449
26. X. Chen, B. Proulx, X. Gong, J. Zhang, Exploiting social ties for cooperative D2D communications: a mobile social networking case. IEEE/ACM Trans. Netw. **23**(5), 1471–1484 (2015)
27. L. Wang, H, Tang, M. Čierny, Device-to-device link admission policy based on social interaction information. IEEE Trans. Veh. Technol. **64**(9), 4180–4186 (2015)
28. L. Wang, G.L. Stüber, Pairing for resource sharing in cellular device-to-device underlays. IEEE Netw. **30**(2), 122–128 (2016)
29. N. Zhang, H. Liang, N. Cheng, Y. Tang, J.W. Mark, X. Shen, Dynamic spectrum access in multi-channel cognitive radio networks. IEEE J. Sel. Areas Commun. **32**(11), 2053–2064 (2014)
30. B. Bai, W. Chen, K.B. Letaief, Z. Cao, A unified matching framework for multi-flow decode-and-forward cooperative networks. IEEE J. Sel. Areas Commun. **30**(2), 397–406 (2012)
31. F.K. Hwang, Control algorithms for rearrangeable Clos networks. IEEE Trans. Commun. **31**(8), 952–954 (1983)

32. T. Sonmez, M.U. Unver, in *Matching, Allocation, and Exchange of Discrete Resources*, Chap. 17, vol. 1A, ed. by J. Benhabib, M.O. Jackson, A. Bisin. Handbook of Social Economics (Elsevier, Amsterdam, 2011), pp. 781–852
33. X. Niu, L. Li, T. Mei, J. Shen, K. Xu, Predicting image popularity in an incomplete social media community by a weighted bi-partite graph, in *Proceedings of IEEE International Conference on Multimedia and Expo* (2012), pp. 735–740
34. L. Wang, H. Wu, W. Wang, K.-C. Chen, Socially enabled wireless networks: resource allocation via bipartite graph matching. IEEE Commun. Mag. **53**(10), 128–135 (2015)
35. D. Feng, L. Lu, Y. Yuan-Wu, G.Y. Li, G. Feng, S. Li, Device-to-device communications underlaying cellular networks. IEEE Trans. Commun. **61**(8), 3541–3551 (2013)

Chapter 3
Proximity Discovery for Cellular D2D Underlay

In D2D enabled networks, direct radio links are allowed between cellular users for data transportation. One of the first steps in setting up D2D links involves pairing UEs that are in close proximity [1]. This is accomplished during neighbor discovery process, where UEs will identify their neighbors for potential direct-link setup. To facilitate the deployment of D2D functionality to standard LTE cellular systems, it is necessary to explore neighbor discovery opportunities in existing LTE infrastructure. In this chapter, we will introduce neighbor discovery techniques for D2D communications of LTE UEs in a modern cellular network [2, 3]. By listening to cellular uplink transmissions, UEs can detect potential D2D partners through a neighbor discovery process compatible with the standard LTE network protocol. We focus on neighbor discovery utilizing the SRS channel, which can be accessed by peer UEs that are LTE-compliant. Under the constraint of unknown channel statistics during uplink hearing, we propose joint neighbor detection and D2D channel estimation for listening UEs using the framework of sparse channel recovery [4–7].

3.1 LTE System Architecture

We consider an LTE cellular network that admits direct user communications. Authenticated UEs in the network have access to the cellular BSs for regular cellular communications. At the same time, UEs can establish D2D links if permitted by the network. To set up D2D communication links, the initiating UE, denoted as UE-0, should learn about its neighbors in advance by listening to neighborhood UEs transmissions. In the cellular network, such UE transmission can be potentially captured during uplink periods. In a scheduled uplink time slot, if UE-0 is not transmitting to the BS, it can listen to other UEs' uplink transmissions and thus

© The Author(s) 2016
L. Wang, H. Tang, *Device-to-Device Communications in Cellular Networks*,
SpringerBriefs in Electrical and Computer Engineering,
DOI 10.1007/978-3-319-30681-0_3

identify UEs with high received signal-to-noise ratio (SNR) as neighbors. UE-0
can initiate D2D communication with neighbor UEs when it needs to transfer data
to them. In the following, we will study neighbor discovery and D2D channel
estimation methods based on cellular uplink hearing.

We will first introduce basic time, frequency, and multiplexing structure of
LTE uplink. Based on this, UE transmissions in different uplink channels are
compared regarding feasibility of implementing neighbor discovery. We identify
SRS and physical random access channel (PRACH) as potential neighbor discovery
opportunities since they are common uplink channels with potential transmissions
from a large number of UEs in the network and it is possible for UE-0 to distinguish
different UE transmissions with available information.

3.1.1   Basic Structure for LTE Uplink

In LTE, multi-user access during uplink is enabled through single-carrier frequency
division multiple access (SC-FDMA). As shown in Fig. 3.1, both discrete fourier
transform (DFT) and inverse discrete Fourier transform (IDFT) are implemented
at the transmitter in a SC-FDMA system to reduce peak-to-average ratio (PAR).
For each OFDM symbol time, the time domain sequence is first transformed into
frequency domain by the DFT block, before being mapped to a set of subcarriers.
Different UEs are mapped to distinct sets of subcarriers, which enables the BS to
separate different UE signals in the frequency domain. After subcarrier mapping,
the zero-filled frequency domain sequence is transformed back to the time domain.
To mitigate inter-symbol interference, cyclic prefix is added to the time domain
sequence before transmission.

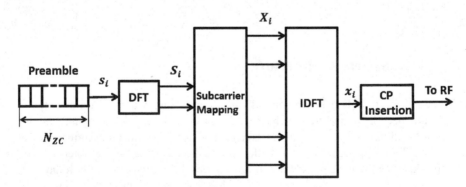

Fig. 3.1 Block diagram of SC-FDMA transmission at UE

Table 3.1 Uplink physical
channels

Channels	Function
PUSCH	Uplink data transmission
PUCCH	Uplink control signal transmission
PRACH	UE random access or reconnection
DM-RS	Channel estimation for coherent demodulation
SRS	Channel sounding for uplink scheduling

3.1.2 Physical Channels for LTE Uplink

The LTE uplink transmissions comprise of three physical channels and two reference signals as listed in Table 3.1. To assess opportunities among these channels where UE-0 can discover its neighbors by eavesdropping on their uplink transmissions, we find it important for UE-0 to have some necessary information on potential transmitters in order to identify different UE transmissions. Since UE-0 does not have knowledge on UE uplink resource allocation, it is difficult to implement neighbor discovery based on signals carried by UE-specific channels. Therefore, we propose to use uplink channels shared by all UEs, such as PRACH and SRS, as potential opportunities for practical D2D neighbor discovery.

(1) *Dedicated Channels:* The physical uplink shared channel (PUSCH) and the physical uplink control channel (PUCCH) are two dedicated channels used for uplink data transportation and control signaling, respectively. The BS assigns different resources to different UEs such that their signal can be separated easily during uplink reception. For PUSCH, distinct sets of resource blocks are allocated to different UEs to avoid conflicts while code division multiplexing (CDM) is used in PUCCH for multi-user access. Since both channels are modulated by UE-specific information, it is difficult for UE-0 to estimate the channel from its neighbors without knowing the specific transmitted data.

(2) *Reference Signals:* There are two reference signals in LTE uplink. One is the demodulation reference signal (DM-RS) located in the middle of PUSCH. As a result, the frequency resources for DM-RS are UE-specific. Unless UE-0 knows fully the resource allocation of other UEs, neighbor discovery based on DM-RS would not be practical. On the other hand, the SRS resides in the last OFDM symbol of each scheduled subframe. Since SRS is used for channel quality estimation to enable frequency-selective scheduling on the uplink, it is transmitted over a large bandwidth to obtain channel information across available subcarriers. Moreover, the subframes used for SRS are broadcasted within the network and known to all UEs. Hence, SRS provides a practical opportunity for neighbor discovery. The detailed discussion of D2D neighbor discovery in SRS is given later.

(3) *Random Access Channel:* The PRACH allows UEs to initiate connection with the BS during its cell entry stage or for reconnection. The PRBs for PRACH are semi-statistically allocated within the PUSCH region and are repeated

periodically. During each PRACH time slot, a transmitting UE may randomly select a preamble from a predefined set to allow BS to distinguish different UE transmissions. Since the preamble set is known to all UEs within the network, UE-0 is able to detect different preamble sequences as part of the neighbor discovery process. Given this feature, we identify PRACH as another potential neighbor discovery opportunity. Due to different channel structure, the neighbor discovery methods in SRS cannot be directly applied to PRACH. However, they can be formulated in the common framework of sparse channel recovery. The solutions for PRACH neighbor discovery and D2D channel estimation will be discussed in a separate work.

3.2 Framework for Neighbor Discovery in LTE

In this section, we present problem formulation of neighbor discovery where a UE listens to neighborhood transmissions during the scheduled SRS symbol. We denote the listening UE as UE-0 and assume that UE-0 listens to SRS channel only when it does not transmit in the corresponding SRS symbol. The proposed framework can be extended to PRACH.

3.2.1 Resource Allocation and Multiplexing

SRS is transmitted on the last SC-FDMA symbol in a subframe. The subframe is claimed by cell-specific broadcast signaling. Data transmission is blocked out in the SRS symbol. UEs are scheduled to transmit in SRS by the BS and they are multiplexed via either frequency division multiplexing (FDM) or CDM. The system bandwidth is divided into disjoint sets of subcarriers. For each subcarrier set (SRS comb [8]), cyclic-shifted Zadoff-Chu (ZC) sequences are used for CDM by up to 8 UEs. The ZC sequence is expressed by

$$s_u(n) = \frac{1}{\sqrt{N}} \exp\left(-j2\pi u \frac{n(n+1)/2}{N}\right), \quad n = 0, \cdots, N-1.$$

u is called the root of the sequence and the sequence length N is an odd number. A ZC sequence has zero correlation with its cyclic-shifted copies and the absolute value of the correlation of two different-root ZC sequences is $\frac{1}{\sqrt{N}}$. Due to this nice property, ZC sequence is commonly adopted in LTE for CDM.

From the multiplexing structure in SRS, if UE-0 wants to distinguish different UEs' transmissions during SRS, it needs to know the allocation of SRS combs and the ZC sequences used for CDM on each SRS comb. In practice, the BS can pass such information to UE-0 through downlink shared channel upon receiving a request for neighbor discovery.

3.2.2 SC-FDMA Transmitter and Receiver

Next we will discuss ZC sequence transmission and reception during SRS. Let $s_i \in C^{n_i}$ denote the ZC sequence used by UE i. n_i is the sequence length. The DFT of s_i is expressed by $S_i = F_{n_i} s_i$, where F_{n_i} denote the DFT matrix of size $n_i \times n_i$. Denote the SRS subcarrier mapping of UE i based on Interleaved FDMA [8] by Γ_i of size $N \times n_i$, then $X_i = \Gamma_i S_i$. N is the total number of subcarriers used for the IDFT $x_i = F_N^T X_i$. Since zero-padding X_i results in an increased sampling rate in the time domain, x_i is an interpolated version of s_i.

Before transmission, cyclic prefix (CP) is prepended to x_i. We assume that the length of CP, denoted as L_{CP}, is larger than the sum of the maximum delay spread L_x and maximum round trip delay (RTD) $\delta_{\max,x}$ in the cell. After adding CP, the transmitted OFDM symbol is $[x_i(N - L_{CP}), \cdots, x_i(N - 1), x_i(0), \cdots, x_i(N - 1)]$. The corresponding signal component $y_i \in C^{(N+L_{CP})}$ received at UE-0 can be represented by

$$
y_{CP,i} = \begin{bmatrix}
x_i(N - L_{CP}) & x_i'(N - 1) & \cdots & x_i'(N - L_x + 1) \\
x_i(N - L_{CP} + 1) & x_i(N - L_{CP}) & \cdots & x_i'(N - L_x + 2) \\
\vdots & & & \vdots \\
x_i(0) & x_i(N - 1) & \cdots & x_i(N - L_x + 1) \\
\vdots & & & \vdots \\
x_i(N - 2) & x_i(N - 3) & \cdots & x_i(N - L_x - 1) \\
x_i(N - 1) & x_i(N - 2) & \cdots & x_i(N - L_x)
\end{bmatrix}
\underbrace{\begin{bmatrix}
h_i(0) \\
h_i(1) \\
\vdots \\
h_i(L_x - 1)
\end{bmatrix}}_{h_i},
$$

where we used $[x_i'(0), \cdots, x_i'(N-1)]$ to denote the OFDM symbol transmitted before SRS. Let N_{SC} denote the total number of SRS combs, then there are at most $U = 8N_{SC}$ SRS transmitters . Due to that some $s_i's$ may not be actively transmitted for a certain SRS opportunity, the actual number of transmitters may be smaller than U. By setting $h_i = 0$ for those inactive ZC sequences, the received signal at UE-0 can be expressed in the following unified form:

$$
y_{CP} = \sum_{i=1}^{U} y_{CP,i}(\delta_{i,x}) + z_{CP}, \tag{3.1}
$$

where $y_{CP,i}(\delta_{i,x})$ denotes the delayed version of $y_{CP,i}$ by $\delta_{i,x}$ and z_{CP} is the additive white Gaussian noise (AWGN). $\delta_{i,x}$ denotes the propagation delay from UE i to UE-0.

From Fig. 3.2, the observation window at UE-0 is of length N and starts at position $L_{CP} - \frac{\delta_{\max,x}}{2}$ of the SRS symbol. (Due to timing advance in LTE system, the earliest possible SRS transmission happens $\frac{\delta_{\max,x}}{2}$ before the SRS symbol.) Then the observed sequence at UE-0 is $y = \sum_{i=1}^{U} y_i + z$ with

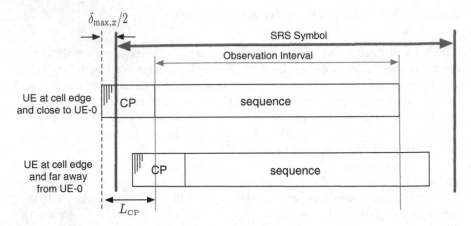

Fig. 3.2 Arrival of ZC sequences at UE-0

$$
y_i = \begin{bmatrix} x_i(N - \delta_{i,x})_N & x_i(N - \delta_{i,x} - 1) & \cdots & x_i(N - \delta_{i,x} - L_x + 1) \\ x_i(N - \delta_{i,x} + 1)_N & x_i(N - \delta_{i,x}) & \cdots & x_i(N - \delta_{i,x} - L_x + 2) \\ \vdots & & & \vdots \\ x_i(N - \delta_{i,x} - 1) & x_i(N - \delta_{i,x} - 2) & \cdots & x_i(N - \delta_{i,x} - L_x) \end{bmatrix} \underbrace{\begin{bmatrix} h_i(0) \\ \vdots \\ h_i(L_x - 1) \end{bmatrix}}_{h_i},
$$

$$(3.2)$$

where $(\cdot)_N$ denotes the modulo-N operation. To separate signals from different SRS combs, UE-0 can demap the received signal in the frequency domain based on SRS comb information acquired from the BS. Let Y, H_i be the DFT of y, h_i, respectively, and denote W_1 as the demapped SRS signal from the SRS comb used by UE-1, then $W_1 = \Gamma_1^T Y$. Indexing the ZC sequences multiplexed on the first comb by 1–8, the kth element of $W_1 \in \mathbb{C}^{n_i}$ is expressed by

$$
W_1(k) = \sum_{i=1}^{8} S_i(k) H_i(k) + Z_1(k). \tag{3.3}
$$

In standard LTE systems, the SRS receiver is designed for the BS. The correlation of the received signal with different cyclic shifts of the root sequence can be used as channel impulse response (CIR) estimates of UEs on the same SRS comb [8]. The frequency domain channel estimates $\hat{H}_i$ are then obtained by applying n_i-point DFT on the CIR estimates. While the BS is able to maintain high SNR through uplink power control, UE-0 will possibly suffer from low receive SNR due to low transmission power or high propagation loss. Moreover, UE-0 does not know which ZC sequences are actively being transmitted. Therefore, active ZC sequence detection becomes a necessary step.

3.2.3 System Model

Before delving into the details of SRS receiver design, we will first present the system model for SRS. In order to take advantage of the correlation property of ZC sequences, transforming Eq. (3.3) to the time domain, we get $w_1 = \sum_{i=1}^{8} r_i + z_1$ where

$$
r_i = \begin{bmatrix} s_i(n_i - \delta_i)_{n_i} & s_i(n_i - \delta_i - 1) & \cdots & s_i(n_i - \delta_i - L + 1) \\ s_i(n_i - \delta_i + 1)_{n_i} & s_i(n_i - \delta_i) & \cdots & s_i(n_i - \delta_i - L + 2) \\ \vdots & & & \vdots \\ s_i(n_i - \delta_i - 1) & s_i(n_i - \delta_i - 2) & \cdots & s_i(n_i - \delta_i - L) \end{bmatrix} \underbrace{\begin{bmatrix} \tilde{h}_i(0) \\ \vdots \\ \tilde{h}_i(L-1) \end{bmatrix}}_{\tilde{h}_i}.
$$

$$(3.4)$$

r_i takes similar form as y_i in Eq. (3.2). Note that x_i and $\tilde{h}_i$ are interpolated versions of s_i and h_i, respectively. Therefore, $\delta_i = \left\lceil \frac{n_i}{N} \delta_{i,x} \right\rceil$, $L = \left\lceil \frac{n_i}{N} L_x \right\rceil$. To overcome the unknown delay of s_i in Eq. (3.4), we consider all possible cases for $\delta_i = 0, \cdots, \delta_{max}$, where $\delta_{max} = \left\lceil \frac{n_i}{N} \delta_{max,x} \right\rceil$. Specifically, let $G = \delta_{max} + L$, r_i can be written as

$$
r_i = \underbrace{\begin{bmatrix} s_i(0) & s_i(n_i - 1) & \cdots & s_i(n_i - G + 1) \\ s_i(1) & s_i(0) & \cdots & s_i(n_i - G + 2) \\ \vdots & & & \vdots \\ s_i(n_i - 1) & s_i(n_i - 2) & \cdots & s_i(n_i - G) \end{bmatrix}}_{\Phi_i} \underbrace{\begin{bmatrix} 0 \\ \tilde{h}_i \\ 0 \end{bmatrix}}_{g_i}.
$$

$$(3.5)$$

$\tilde{h}_i$ starts at the $(\delta_i + 1)$-th position in g_i. Based on Eq. (3.5), we get $w_1 = \sum_{i=1}^{8} \Phi_i g_i + z_1$. Applying the same transformation across N_{SC} SRS combs, the jointly received signal $w := [w_1^T, \cdots, w_{N_{sc}}^T]^T$ can be expressed by

$$
w = \underbrace{\begin{bmatrix} \Phi_1 \cdots \Phi_8 & 0 & \cdots & 0 \\ 0 & \ddots & & 0 \\ 0 & & \ddots & 0 \\ 0 & \cdots & 0 & \Phi_{U-7} \cdots \Phi_U \end{bmatrix}}_{\Psi} \underbrace{\begin{bmatrix} g_1 \\ \vdots \\ \vdots \\ g_U \end{bmatrix}}_{g} + \underbrace{\begin{bmatrix} z_1 \\ \vdots \\ \vdots \\ z_U \end{bmatrix}}_{z}.
$$

$$(3.6)$$

$\Psi \in \mathbb{C}^{M_r \times M_c}$ with $M_r = \frac{1}{8} \sum_{i=1}^{U} n_i$ and $M_c = UG$. Due to that the columns of $\Phi_i's$ on the same row are ZC sequences generated from the same root, $\Psi^H \Psi = I_{M_c}$. Since each root sequence provides CDM for eight users, it is assumed that $n_i > 8G$ for all i. Consequently, Ψ is a tall matrix with full column rank. Denote the columns of Ψ containing Φ_i as Ψ_i, we can represent w by $w = \sum_{i=1}^{U} \Psi_i g_i + z = \Psi g + z$.

3.2.4 Sparse Vector Recovery

Based on the system model in Eq. (3.6), our main purpose is to extract the
stacked channel vector g from the noisy observation w, given the knowledge of
the coefficient matrix $\boldsymbol{\Psi}$. g is a block sparse vector [9] that admits block structure
corresponding to each UE. Assuming the channel vectors and the noise vector
follow independent complex Gaussian distribution, $z \sim \text{CN}(0, \lambda I_{M_r})$ where λ is
the noise variance, $g_i \sim \text{CN}(0, \Sigma_{g_i})$ where Σ_{g_i} has the structure

$$\Sigma_{g_i} = \begin{bmatrix} \mathbf{0} & & \\ & \gamma_i I_{L_i} & \\ & & \mathbf{0} \end{bmatrix}_{G \times G}. \tag{3.7}$$

γ_i and L_i are the channel variance and delay spread of UE i, respectively. $\gamma_i = 0$
for the ZC sequences that are inactive. The nonzero diagonal entries in Σ_{g_i}
starts at the $(\delta_i + 1)$-th position corresponding to the position of $\tilde{h}_i$ in g_i. From
Eq. (3.6), $g \sim \text{CN}(0, \Sigma_g)$ with $\Sigma_g = \text{diag}(\Sigma_{g_1}, \Sigma_{g_2}, \cdots, \Sigma_{g_U})$. Using the statistical
model above, recovering g can be interpreted as a pure estimation problem of the
parameter set $\boldsymbol{\theta} = \left\{ \{\gamma_i, \delta_i, L_i\}_{i=1}^{U}, \lambda \right\}$. By adopting the framework of block sparse
Bayesian learning (BSBL), g can be recovered using maximum a posteriori (MAP)
criterion based on maximum likelihood (ML) estimate of $\boldsymbol{\theta}$. $\hat{\gamma}_i = 0$ indicates the
corresponding s_i is not active.

3.3 Block Sparse Bayesian Learning

In this section, we will focus on using BSBL proposed in [9, 10] for sparse vector
recovery given the system model Eq. (3.6). While uniform covariance structure is
assumed in [9, 10] for each block, we are faced with the uncertainty of δ_i and L_i
of each UE. However, by taking advantage of the orthogonality of $\boldsymbol{\Psi}_i$, $\boldsymbol{\theta}$ can be
estimated using ML estimation in the BSBL framework.

3.3.1 Problem Formulation

In Eq. (3.6), given the statistical model of g and z, the distribution of w is
$w \sim \text{CN}(0, \Sigma_w)$ with $\Sigma_w = \boldsymbol{\Psi} \Sigma_g \boldsymbol{\Psi}^H + \lambda I_{M_r}$. With the observed vector w, the
posterior density of g also follows Gaussian distribution denoted by $p(g|w, \Sigma_g) \sim$
$\text{CN}(u_{\text{MAP}}, \Sigma_{\text{MAP}})$ with

$$u_{\text{MAP}} = \Sigma_g \boldsymbol{\Psi}^H (\boldsymbol{\Psi} \Sigma_g \boldsymbol{\Psi}^H + \lambda I_{M_r})^{-1} w, \tag{3.8a}$$

$$\Sigma_{\text{MAP}} = \Sigma_g - \Sigma_g \boldsymbol{\Psi}^H (\boldsymbol{\Psi} \Sigma_g \boldsymbol{\Psi}^H + \lambda I_{M_r})^{-1} \boldsymbol{\Psi} \Sigma_g. \tag{3.8b}$$

The essential idea of Bayesian learning is that the observation w can provide additional evidence to refine the estimate of g, which is performed by maximizing $p(g|w, \Sigma_g)$. For known θ, the MAP estimate of g can be derived directly as

$$\hat{g} = u_{\text{MAP}}. \tag{3.9}$$

For u_{MAP} with unknown parameters, the critical issue of Bayesian learning lies in parameter estimation. Using ML criterion, the parameter estimation in BSBL is expressed by

$$\underset{\theta}{\text{Maximize}} \quad \log p(w \mid \theta)$$

which, after substituting the distribution of w, is equivalent to

$$\underset{\theta}{\text{Minimize}} \quad \xi_S = \log |\Sigma_w| + w^H \Sigma_w^{-1} w. \tag{3.10}$$

3.3.2 Maximum Likelihood Estimation

Next we discuss solving Eq. (3.10) for the neighbor discovery model in Eq. (3.6). Considering the block structure of Σ_g and that Ψ has orthonormal columns, we can decompose Eq. (3.10) with respect to each block. First, denote

$$\overline{\Psi} = [\Psi \ \Psi_{U+1}],$$

where Ψ_{U+1} is the null space of Ψ^H with $K = M_r - M_c$ orthonormal columns. Correspondingly, Σ_g can be expanded into diagonal matrix $\overline{\Sigma}_g$ by padding K zeros on the diagonal. Substitute $\overline{\Psi}$ and $\overline{\Sigma}_g$ into Eq. (3.10), we get

$$\xi_S = \log \left| \Sigma_g \Psi^H \Psi + \lambda I_{M_c} \right| + K \log \lambda + w^H \overline{\Psi} \left(\overline{\Sigma}_g + \lambda I_{M_r} \right)^{-1} \overline{\Psi}^H w. \tag{3.11}$$

Let $v_i := \Psi_i^H w = g_i + \Psi_i^H z$ for $i = 1, \cdots, U+1$, we can decompose Eq. (3.11) as

$$\xi_S = \sum_{i=1}^{U} \left[\log \left| \Sigma_{g_i} + \lambda I_G \right| + v_i^H \left(\Sigma_{g_i} + \lambda I_G \right)^{-1} v_i \right] + K \log \lambda + \frac{\|v_{U+1}\|_2^2}{\lambda}. \tag{3.12}$$

The terms in the square bracket can be further transformed as

$$\xi_S^{(i)} := L_i \log(\lambda + \gamma_i) + (G - L_i) \log \lambda + \frac{\|v_{i,L_i}\|_2^2}{\lambda + \gamma_i} + \frac{\|v_{i,L_i^c}\|_2^2}{\lambda},$$

where $\boldsymbol{v}_{i,L_i}$ denotes the L_i consecutive elements within $\boldsymbol{v}_i$ starting at the $(\delta_i + 1)$-th position, whereas $\boldsymbol{v}_{i,L_i^c}$ denotes the remaining entries. Assign $\xi_S^{(U+1)} := K \log \lambda + \frac{\|\boldsymbol{v}_{U+1}\|_2^2}{\lambda}$. Then $\xi_S = \sum_{i=1}^{U+1} \xi_S^{(i)}$.

To minimize ξ_S, we first note that for $\gamma_i \geq 0$, $\xi_S^{(i)}$ is minimized when $\boldsymbol{v}_{i,L_i}$ is taken as the L_i consecutive elements in $\boldsymbol{v}_i$ with the largest l_2-norm (energy), which gives the estimate of δ_i. To estimate $\{\gamma_i\}_{i=1}^{U}$, we can start by extracting the terms related to γ_i, which we denote as

$$f_{\gamma_i}(\gamma_i) = L_i \log(\gamma_i + \lambda) + \frac{\|\boldsymbol{v}_{i,L_i}\|_2^2}{\gamma_i + \lambda}.$$

For given L_i and λ, the minimizer of $f_{\gamma_i}(\gamma_i)$ is given by

$$\hat{\gamma}_i = \max\left(\frac{\|\boldsymbol{v}_{i,L_i}\|_2^2}{L_i} - \lambda, 0\right) =: \left(\frac{\|\boldsymbol{v}_{i,L_i}\|_2^2}{L_i} - \lambda\right)^+. \tag{3.13}$$

Correspondingly,

$$f_{\gamma_i}(\hat{\gamma}_i) = \begin{cases} L_i \log \lambda + \frac{\|\boldsymbol{v}_{i,L_i}\|_2^2}{\lambda} , & \lambda > \frac{\|\boldsymbol{v}_{i,L_i}\|_2^2}{L_i} \\ L_i \log \frac{\|\boldsymbol{v}_{i,L_i}\|_2^2}{L_i} + L_i , & \text{otherwise} \end{cases}. \tag{3.14}$$

Substituting $\hat{\gamma}_i$ into $\xi_S^{(i)}$, we get

$$\xi_S^{(i)} = \begin{cases} G \log \lambda + \frac{\|\boldsymbol{v}_i\|_2^2}{\lambda}, & \lambda > \frac{\|\boldsymbol{v}_{i,L_i}\|_2^2}{L_i} \\ L_i \log \frac{\|\boldsymbol{v}_{i,L_i}\|_2^2}{L_i\lambda} + L_i + G \log \lambda + \frac{\|\boldsymbol{v}_{i,L_i^c}\|_2^2}{\lambda}, & \text{otherwise} \end{cases}. \tag{3.15}$$

Since the second expression in Eq. (3.15) is no larger than the first expression, L_i is estimated as

$$\hat{L}_i = \underset{\substack{L_i \in \{1, \cdots, L\} \\ \|\boldsymbol{v}_{i,L_i}\|_2^2/L_i \geq \lambda}}{\arg\min} \left\{ L_i \log \frac{\|\boldsymbol{v}_{i,L_i}\|_2^2}{L_i\lambda} + L_i + \frac{\|\boldsymbol{v}_{i,L_i^c}\|_2^2}{\lambda} \right\}. \tag{3.16}$$

Note that $\max_{L_i} \frac{\|\boldsymbol{v}_{i,L_i}\|_2^2}{L_i} = \max_j |v_{i,j}|^2$, where $v_{i,j}$ is the j-th element of $\boldsymbol{v}_i$. From Eq. (3.15), when $\max_j |v_{i,j}|^2 < \lambda$, we can take $\hat{L}_i = 0$ and $\hat{\gamma}_i = 0$. Denote

$$\mathrm{I} = \left\{ i : \max_j |v_{ij}|^2 \geq \lambda \right\}. \tag{3.17}$$

Substituting Eq. (3.15) into Eq. (3.12), we have

$$\xi_S = \sum_{i \in I} \left[\hat{L}_i \log \frac{\|v_{i,\hat{L}_i}\|_2^2}{\hat{L}_i \lambda} + \hat{L}_i + G \log \lambda + \frac{\|v_{i,\hat{L}_i^c}\|_2^2}{\lambda} \right]$$

$$+ \sum_{i \notin I} \left[G \log \lambda + \frac{\|v_i\|_2^2}{\lambda} \right] + K \log \lambda + \frac{\|v_{U+1}\|_2^2}{\lambda},$$

whose minimizer is given by

$$\hat{\lambda} = \frac{\sum_{i=1}^{U+1} \|v_i\|_2^2 - \sum_{i \in I} \|v_{i,\hat{L}_i}\|_2^2}{UG + K - \sum_{i \in I} \hat{L}_i}. \tag{3.18}$$

Since $\hat{\lambda}$, I, and $\hat{L}_i$ are mutually dependent, we update the estimates iteratively starting with $\hat{\lambda}^{(0)}$ and then applying Eqs. (3.16), (3.17), and (3.18) alternatively until ξ_S no longer decreases. Since both Eqs. (3.16) and (3.18) reduce ξ_S and ξ_S is lower bounded by

$$\check{\xi}_S = \sum_{i=1}^{U} \check{\xi}_S^{(i)} + K \log \frac{\|v_{U+1}\|_2^2}{K} + K, \tag{3.19}$$

where $\check{\xi}_S^{(i)}$ is the lower bound of $\xi_S^{(i)}$ in Eq. (3.15) given by

$$\check{\xi}_S^{(i)} = \min_{L_i} \left\{ L_i \log \frac{\|v_{i,L_i}\|_2^2}{L_i} + (G - L_i) \log \frac{\|v_{i,L_i^c}\|_2^2}{G - L_i} + G \right\}, \tag{3.20}$$

the iteration will converge. The convergence point depends on $\hat{\lambda}^{(0)}$, which can be estimated from the noise subspace as

$$\hat{\lambda}^{(0)} = \frac{\|v_{U+1}\|_2^2}{K}. \tag{3.21}$$

Note that $v_i = \Psi_i w$ is the correlation of the received signal with different ZC sequences. Therefore, BSBL can be applied to the correlation-based SRS receiver designed for the BS [8].

3.3.3 Simulation Results and Performance Analysis

In this section, we will provide numerical results to evaluate the performances of block sparse Bayesian learning method for neighbor discovery.

Table 3.2 Simulation parameters

D2D path loss	148 + 40*log10(d[km])
Uplink path loss	128.1 + 37.6*log10(d[km])
Noise variance [dB]	−174 + 10*log10(BW)+NF-30
Noise figure (NF)	7
Bandwidth (BW)	20 MHz (100 RB)
DFT size	2048
SRS bandwidth	96 RB (1152 subcarriers)
ZC sequence length	571
Cyclic prefix duration	160 Ts (5.2 us)
Maximum delay spread	0.5 us (ETU model)
Cell radius	600 m
Maximum round trip delay	4 us

Table 3.3 Size parameters

x_i	$N = 2048$	$L_{CP} = 160$	$L_x = 16$	$\delta_{max,x} = 124$
s_i	$n_i = 571$	$G = 40$	$L = 5$	$\delta_{max} = 35$
	Cyclic shift $= 71$		$U = 16$	$K = 502$

3.3.3.1 Simulation Parameters

The channel parameters following LTE standards [2, 3] are provided in Table 3.2. For the simulation results presented in the paper, we considered two SRS combs and hence in total 16 possible SRS transmitters. Due to IFDMA, each SRS comb is composed of 576 subcarriers, which is half of the SRS bandwidth. In Table 3.3, x_i denotes the parameters in Eq. (3.2), which is based on 2048-point DFT assumed for uplink SC-FDMA. s_i denotes the corresponding size parameters in Eq. (3.4) for a SRS comb. L_{CP}, L_x, and $\delta_{max,x}$ are calculated based on the CP duration, maximum delay spread, and maximum RTD in Table 3.2.

In LTE, the transmission power of each UE is adjusted such that each uplink maintains a fixed SNR at the BS. From the path loss model in Table 3.2, UEs located further from the BS will adopt higher transmission power. Denote the target uplink SNR as $SNR_T[dB]$. The transmission power [dB] of UE i with uplink path loss PL_i is calculated as

$$P_{T,i}[dB] = SNR_T[dB] + PL_i[dB] + \sigma_n^2[dB].$$

PL_i and the noise variance σ_n^2 are listed in Table 3.2. We consider neighbor discovery process within one cell where 10 out of 16 possible SRS transmitters are actively transmitting. The locations of the transmitters are uniformly generated over an area of 600 × 600 m square centered at the BS. UE-0 is assumed 300 m away from the BS. The channel vectors and the noise vectors are generated randomly following the complex Gaussian model given in Section III. Each simulation consists of 1000 trials.

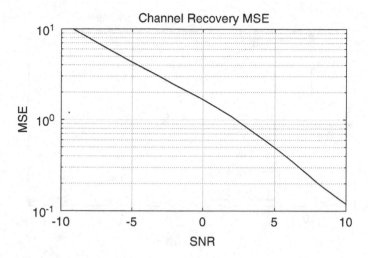

Fig. 3.3 MSE of sparse channel recovery

3.3.3.2 Performance of Sparse Channel Recovery

We plot the active ZC sequence detection probability and stacked channel vector mean square error (MSE) as a function of SNR in Fig. 3.3. The channel MSE is calculated by $\|g - \hat{g}\|_2^2/(\sum_{i=1}^{U} L_i\gamma_i)$, which captures both detection error and estimation error of all the parameters. Naturally, higher transmission power leads to better performance of the system.

3.3.3.3 Error Probability of Parameter Estimation

In Fig. 3.4, we evaluate the performance of δ_i estimation, which is the propagation delay of the i-th transmitter. We study the results under different multi-antenna configuration, where N_a indicates the number of transmitting antennas. From the figure, we can see that the error probability drops with SNR. Furthermore, more transmitting antennas increase the possibility of accurate estimate of the parameter.

3.3.3.4 Detection Probability

To show the effectiveness of block sparse Bayesian algorithm for neighbor discovery, we plot the number of detected transmitters as a function of the total number of transmitters in Fig. 3.5. The figure shows that the transmitters can be discovered with a high probability.

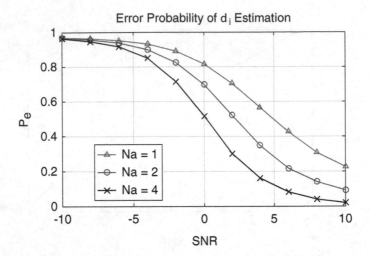

Fig. 3.4 Detection error probability

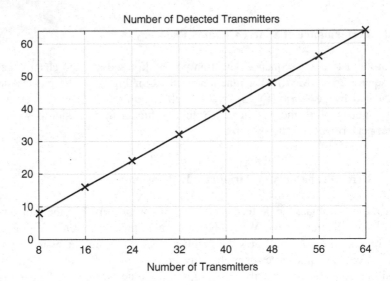

Fig. 3.5 Number of detected transmitters

3.4   Chapter Summary

In this chapter, we focus on the problem of neighbor discovery for enabling direct
UE communications. It is demonstrated that users in the cellular network can detect
their neighbors by simply listening to their uplink transmissions. We compare

different uplink channels in the LTE system and propose SRS and PRACH as potential neighbor discovery opportunities due to their common LTE structure to all the users.

Our work focuses on statistical methods for simultaneous neighbor detection and D2D channel estimation. Due to the fact that the channel parameters are unknown to the discovering UE, we propose several methods for neighbor discovery in SRS using the framework of block sparse Bayesian learning and composite hypothesis testing in detection theory. The estimated D2D channel during neighbor discovery process can be utilized for later D2D communications. The neighbor discovery performance is evaluated with regards to various system parameters in practical LTE deployment.

References

1. Feasibility Study for Proximity Services (ProSe) (Release 12). 3GPP TSG SA TR 22.803, v 1.0.0, August 2012
2. S. Sesia, I. Toufik, M. Baker, *LTE: The UMTS Long Term Evolution* (Wiley, New York, 2009)
3. Evolved Universal Terrestrial Radio Access (E-UTRA); Physical channels and modulation (Release 11), 3GPP TSG RAN TS 36.211, v 11.3.0, July 2013
4. D. Angelosante, E. Biglieri, M. Lops, Neighbor discovery in wireless networks: a multiuser-detection approach. Phys. Commun. 3(1), 28–36 (2010)
5. H. Zhu, G.B. Giannakis, Exploiting sparse user activity in multiuser detection. IEEE Trans. Commun. 59(2), 454–465 (2011)
6. C. Bockelmann, H.F. Schepker, A. Dekorsy, Compressive sensing based multi-user detection for machine-to-machine communication. Trans. Emerg. Telecommun. Technol. 24(4), 389–400 (2013)
7. A.K. Fletcher, S. Rangan, V.K. Goyal, On-off random access channels: a compressed sensing framework (2009). Available: http://arxiv.org/abs/0903.1022
8. P. Bertrand, Channel gain estimation from sounding reference signal in LTE, in *IEEE 73rd Vehicular Technology Conference (VTC Spring)*, May 2011
9. Z. Zhang, B. Rao, Sparse signal recovery with temporally correlated source vectors using sparse Bayesian learning. IEEE J. Sel. Top. Sign. Process. 5(5), 912–926 (2011)
10. Z. Zhang, B. Rao, Extension of SBL algorithms for the recovery of block sparse signals with intra-block correlation. IEEE Trans. Signal Process. 61(8), 2009–2015 (2013)

Chapter 4
Mode Selection for Cellular D2D Underlay

In cellular communication systems with optional D2D links, UEs can operate in either D2D mode or cellular mode for data transport [1–3]. In this chapter, we introduce mixed-mode D2D communication in which D2D links can operate in multiple modes through resource multiplexing. We study the problem of maximizing weighted D2D sum rate under cellular rate constraints by optimizing mixed-mode allocation and resource allocation in term of transmit power and subchannel assignment. A two-step approach is proposed by introducing energy-splitting variables such that mixed-mode allocation and resource allocation can be decoupled and optimized independently. The resulting algorithm can be distributive, requires little signaling overhead and has low computational complexity.

4.1 System Model

We consider an LTE cellular network in which CUEs in each cell are assigned orthogonal resource block (RB) chunks for either uplink or downlink communications with the BS. As illustrated in Fig. 4.1, the resource pool of the cell contains the uplink and downlink RB chunks of CUEs that are in steady communication status and have low data rate requirements. Let $\mathscr{C} = \{1, \cdots, J\}$ be the set of CUEs. The resource pool can be denoted as $\mathscr{R} = \{1, 2, \cdots, 2j-1, 2j, \cdots, K\}$, where $2j-1$ and $2j$ denote the uplink and downlink RB chunk of CUE j, respectively, and $K = 2J$. This notation applies to both frequency division duplex (FDD) and time division duplex (TDD) cellular systems since they both guarantee non-overlapping resource allocation among CUEs. Also, we consider common channel state across each RB chunk by selecting a small enough number of resource blocks for each chunk.

© The Author(s) 2016

L. Wang, H. Tang, *Device-to-Device Communications in Cellular Networks*,
SpringerBriefs in Electrical and Computer Engineering,
DOI 10.1007/978-3-319-30681-0_4

Fig. 4.1 Resource pool

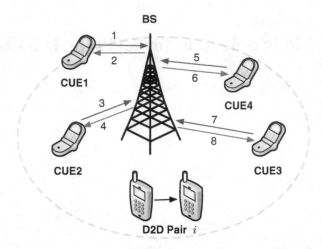

4.1.1   Channel Model

To limit interference from D2D communications to the cellular system, we assume that each RB chunk can be shared by at most one D2D link. Denote $\mathcal{D} = \{1, \cdots, I\}$ as the set of D2D links. For a fixed pairing (i, k) between D2D links and RB chunks, the UEs and the BS can optionally operate in one of the following modes:

$$m = \begin{cases} 1, \text{ Pure Cellular} \\ 2, \text{ D-B-D} \\ 3, \text{ Dedicated D2D} \\ 4, \text{ Underlay D2D} \end{cases}, \tag{4.1}$$

where m is the mode indicator. $m = 1$ represents regular cellular communication without D2D communication. $m = 2$ indicates D-B-D (Device-BS-Device) mode where the D2D devices communicate through the BS without co-channel cellular transmission. Figure 4.2a depicts the scenario. $m = 3$ represents dedicated D2D communication where the D2D devices communicate directly without co-channel cellular transmission. In Underlay D2D mode ($m = 4$), the direct link and the cellular link transmit simultaneously as illustrated in Fig. 4.2b. We denote M = $\{1, \cdots, 4\}$ as the collection of all the modes. Furthermore, the tuple (i, k, m) is used as subscript throughout the paper to indicate variables in mode m when RB chunk k is allocated to D2D link i. Let $r_{ikm}^{(C)}$ and $r_{ikm}^{(D)}$ denote the cellular and D2D link rate, respectively. Depending on whether RB chunk k is cellular uplink or downlink resource, the rate functions differ as given in Table 4.1, where

$$r_{ikm}^{(UL)} = 0.5 \log \left(1 + p_{ikm} \left| \boldsymbol{v}_{ikm}^{H} \boldsymbol{h}_{ie}^{(k)} \right|^2 \right) \tag{4.2}$$

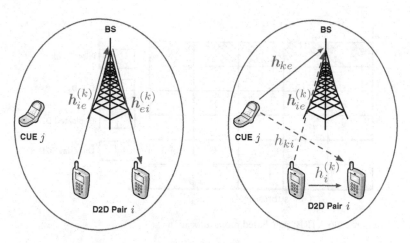

Fig. 4.2 (**a**) D-B-D mode and (**b**) Underlay D2D mode

Table 4.1 Rate functions

Uplink sharing		
m	$r_{ikm}^{(C)}$	$r_{ikm}^{(D)}$
1	$\log\left(1 + P_k \left| v_{ikm}^H h_{ke} \right|^2 \right)$	0
2	0	$\min\left(r_{ikm}^{(UL)}, r_{ikm}^{(DL)}\right)$
3	0	$\log\left(1 + p_{ikm}\left| h_i^{(k)} \right|^2 \right)$
4	$\log\left(1 + \dfrac{p_{ikm}\left| h_i^{(k)} \right|^2}{P_k |h_{ki}|^2 + 1}\right)$	$\log\left(1 + \dfrac{P_k |v_{ikm}^H h_{ke}|^2}{p_{ikm}|v_{ikm}^H h_{ie}^{(k)}|^2 + 1}\right)$
Downlink sharing		
m	$r_{ikm}^{(C)}$	$r_{ikm}^{(D)}$
1	$\log\left(1 + P_e \left| h_{ek}^H w_{ikm} \right|^2 \right)$	0
2	0	$\min\left(r_{ikm}^{(UL)}, r_{ikm}^{(DL)}\right)$
3	0	$\log\left(1 + p_{ikm}\left| h_i^{(k)} \right|^2 \right)$
4	$\log\left(1 + \dfrac{p_{ikm}\left| h_i^{(k)} \right|^2}{P_e \left|\left(h_{ei}^{(k)}\right)^H w_{ikm}\right|^2 + 1}\right)$	$\log\left(1 + \dfrac{P_e |h_{ek}^H w_{ikm}|^2}{p_{ikm}|h_{ik}|^2 + 1}\right)$

$$r_{ikm}^{(DL)} = 0.5\log\left(1 + P_e \left|\left(h_{ei}^{(k)}\right)^H w_{ikm}\right|^2\right) \tag{4.3}$$

are the rate functions of uplink hop and downlink hop in the D-B-D mode. The channel vectors in Table 4.1 are denoted in Fig. 4.2. w_{ikm} and v_{ikm} are the normalized transmit and receive precoder of the BS. Without loss of generality, we assume unit noise variance. p_{ikm} is the transmission power of D2D link i on RB chunk

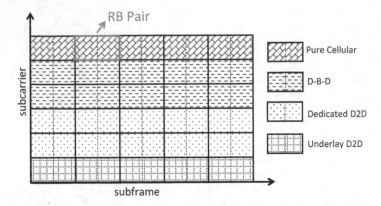

Fig. 4.3 Resource block (RB) pair-based mode allocation

k in mode m, while P_k and P_e denote the transmission power of the CUE on RB chunk k and the transmission power of the BS, respectively. We focus on optimizing D2D communication parameters and assume minimum disruption to existing cellular infrastructure. Therefore, P_k and P_e are assumed constants that are specified according to the standard cellular power control protocol [4].

4.1.2 Resource Multiplexing

One critical issue of enabling mixed-mode D2D operation concerns its practical implementations in existing LTE systems. This problem can be resolved by segmenting each RB chunk such that distinct RBs are used for different modes. In LTE systems, the radio resources are allocated to UEs in units of RB pairs. As illustrated in Fig. 4.3, the RB pair is a two dimensional block that consists of one subframe in the time domain and 12 subcarriers in the frequency domain. Each subframe has a duration of 1ms. Let N_k be the number of RB pairs in RB chunk k and x_{ikm} be the fraction of RB chunk k allocated to D2D link i for mode m. Then $\lceil N_k x_{ikm} \rceil$ RB pairs will be used by the BS and corresponding UEs to operate in mode m. The mode allocation can be implemented through FDM, time division multiplexing (TDM) or across the two dimensional plane. Figure 4.3 illustrates FDM as an example, where different modes occupy different frequency bands within the RB chunk. The RB pair-based mode allocation can be implemented effectively without additional signaling overhead by devising simple labeling protocols. Take Fig. 4.3 as an example. By indexing the subcarriers and allocating mode continuously, the D2D and cellular transmitter can simultaneously locate the starting subcarrier of each mode and operate in the corresponding mode.

4.2   Problem Formulation

To address the multi-objective nature of D2D underlay system, we formulate the joint mode and resource allocation problem with the objective of maximizing weighted D2D sum rate under cellular rate constraints. We use μ_{ik} as the pairing indicator between D2D link i and RB chunk k, which takes binary value as below

$$\mu_{ik} = \begin{cases} 1, & \text{if RB chunk } k \text{ is reused by D2D link } i \\ 0, & \text{otherwise} \end{cases}. \tag{4.4}$$

Further, $\beta_i \geq 0$ is the weight of D2D link i, which can be specified according to system fairness and priority requirements [5]. Define $\{x_{ikm}, p_{ikm}, w_{ikm}, v_{ikm}\}$ and $\{\mu_{ik}\}$ as the collection of variables over all the index sets. The mixed-mode resource allocation problem is formulated as

$$\underset{\substack{\{x_{ikm}, p_{ikm}\} \\ \{w_{ikm}, v_{ikm}\} \\ \{\mu_{ik}\}}}{\text{maximize}} \quad \sum_{i=1}^{I} \beta_i \sum_{k=1}^{K} \mu_{ik} \sum_{m=1}^{4} x_{ikm} r_{ikm}^{(D)} \tag{4.5a}$$

$$\text{subject to} \quad \sum_{i=1}^{I} \mu_{ik} \sum_{m=1}^{4} x_{ikm} r_{ikm}^{(C)} \geq R_k \quad \forall k \in \mathcal{R} \tag{4.5b}$$

$$\sum_{k=1}^{K} \mu_{ik} \sum_{m=1}^{4} x_{ikm} p_{ikm} \leq E_i \quad \forall i \in \mathcal{D} \tag{4.5c}$$

$$\sum_{m=1}^{4} x_{ikm} \leq 1, \quad x_{ikm} \geq 0 \quad \forall i \in \mathcal{D}, k \in \mathcal{R} \tag{4.5d}$$

$$\sum_{i=1}^{I} \mu_{ik} \leq 1, \quad \mu_{i,k} \in \{0, 1\} \quad \forall k \in \mathcal{R} \tag{4.5e}$$

In the problem, Eq. (4.5b) enforces that the rate of each cellular link must be higher than a given threshold R_k, which characterizes the minimum QoS requirements of prioritized CUEs. For feasibility concern, it is assumed that R_k is smaller than the cellular rate in Pure Cellular Mode. Equation (4.5c) is the sum energy constraint of each D2D transmitter. Equation (4.5d) requires that the sum of all radio resource fractions in different modes cannot be larger than one. Equation (4.5e) regulates that each RB chunk can only be reused by at most one D2D link.

Due to the binary constraints of $\{\mu_{ik}\}$, problem Eq. (4.5) is a combinatorial problem with I^K possible pairings between D2D links and RB chunks. The computational complexity is prohibitively high in cellular systems where the number of RB chunks and the number of D2D links are usually very large. Moreover, the cellular rate constraints Eq. (4.5b) are non-convex, which makes problem Eq. (4.5) generally intractable.

4.2.1 Two-Step Approach

To tackle the issues of non-convexity and combinatorial integer constraints, we approach problem Eq. (4.5) by first investigating the optimal mode allocation for a fixed pairing (i, k) between D2D links and RB chunks. The mode allocation problem for fixed (i, k) is expressed by

$$Q_{ik}(E_{ik}) \triangleq \underset{\{x_m, p_m\} \atop \{w_m, v_m\}}{\text{maximize}} \quad \sum_{m=1}^{4} x_m r_m^{(D)} \tag{4.6a}$$

$$\text{subject to} \quad \sum_{m=1}^{4} x_m r_m^{(C)} \geq R_k \tag{4.6b}$$

$$\sum_{m=1}^{4} x_m p_m \leq E_{ik} \tag{4.6c}$$

$$\sum_{m=1}^{4} x_m \leq 1, \quad x_m \geq 0 \tag{4.6d}$$

Noting that problem Eq. (4.6) is optimized for every pairing (i, k), we omit indices i, k in the rate functions and optimization variables above for simplicity. Constraint Eq. (4.6b) results from the cellular rate constraints Eq. (4.5b). Since each RB chunk can be reused by at most one D2D link, the cellular rate must be larger than the corresponding threshold R_k for a valid pairing (i, k). Considering that each D2D link may possibly be allocated multiple RB chunks, we handle the sum energy constraint Eq. (4.5c) by introducing *energy-splitting variables* E_{ik}. $E_{ik} \geq 0$ and $\sum_{k=1}^{K} \mu_{ik} E_{ik} \leq E_i$. In problem Eq. (4.6), E_{ik} can be taken as a constant. Nevertheless, the objective value Eq. (4.6a) will be a function of E_{ik}, which we denoted as $Q_{ik}(E_{ik})$. Substituting $Q_{ik}(E_{ik})$, problem Eq. (4.5) becomes

$$\underset{\{\mu_{ik}, E_{ik}\}}{\text{maximize}} \quad \sum_{i=1}^{I} \beta_i \sum_{k=1}^{K} \mu_{ik} Q_{ik}(E_{ik}) \tag{4.7a}$$

$$\text{subject to} \quad \sum_{k=1}^{K} \mu_{ik} E_{ik} \leq E_i, \quad E_{ik} \geq 0 \qquad \forall i \in \mathscr{D} \tag{4.7b}$$

$$\sum_{i=1}^{I} \mu_{ik} \leq 1, \quad \mu_{ik} \in \{0, 1\} \qquad \forall k \in \mathscr{R} \tag{4.7c}$$

Problem in Eq. (4.7) is a standard joint subcarrier and power allocation problem for cellular uplink [5]. However, the solution of the problem depends on the function $Q_{ik}(E_{ik})$. In the following sections, we will first study problem Eq. (4.6) to characterize $Q_{ik}(E_{ik})$. Based on the results, problem Eq. (4.7) is further discussed with practical implementation considerations.

4.3 Mixed-Mode Allocation

In this section, we will discuss the solution to the mode allocation problem Eq. (4.6). We first focus on optimizing the multi-antenna vectors w_m and v_m for uplink and downlink sharing, respectively. Based on the optimal multi-antenna vectors, problem Eq. (4.6) is simplified with only optimization variables $\{x_m, p_m\}$, which is solved by successive convex approximation method.

4.3.1 Uplink Sharing

In this subsection, we study the optimal multi-antenna vectors of problem Eq. (4.6) for uplink sharing scenario. For ease of discussion, we summarize the expressions of D2D rate functions in uplink sharing scenario as below

$$
r_m^{(D)} = \begin{cases} 0, & \text{Pure Cellular} \\ \min\left(r_2^{(UL)}, r_2^{(DL)}\right), & \text{D-B-D} \\ \log\left(1 + p_3 \left|h_i^{(k)}\right|^2\right), & \text{Dedicated D2D} \\ \log\left(1 + \dfrac{p_4 \left|h_i^{(k)}\right|^2}{P_k |h_{ki}|^2 + 1}\right), & \text{Underlay D2D} \end{cases} \tag{4.8}
$$

where

$$
r_2^{(UL)} = 0.5 \log\left(1 + p_2 \left|v_2^H h_{ie}^{(k)}\right|^2\right), \tag{4.9}
$$

$$
r_2^{(DL)} = 0.5 \log\left(1 + P_e \left|\left(h_{ei}^{(k)}\right)^H w_2\right|^2\right). \tag{4.10}
$$

We will show that $r_2^{(D)} = r_2^{(UL)}$ for the optimal solution of problem Eq. (4.6) using proof by contradiction. Suppose $r_2^{(UL)} > r_2^{(D)}$ for the optimal solution $\{x_m^*, p_m^*, w_m^*, v_m^*\}$ of problem Eq. (4.6). Let $\{\tilde{x}_m, \tilde{p}_m, \tilde{w}_m, \tilde{v}_m\} = \{x_m^*, p_m^*, w_m^*, v_m^*\}$ except that $\tilde{p}_2 = p_2^* - \epsilon$ and $\tilde{p}_3 = p_3^* + \epsilon$, where $\epsilon > 0$ is chosen such that $r_2^{(UL)} = r_2^{(D)}$. Then $\{\tilde{x}_m, \tilde{p}_m, \tilde{w}_m, \tilde{v}_m\}$ leads to larger objective value in Eq. (4.6a) than $\{x_m^*, p_m^*, w_m^*, v_m^*\}$ due to increment of $r_3^{(D)}$. This contradicts with the fact that $\{x_m^*, p_m^*, w_m^*, v_m^*\}$ is optimal. Therefore, $r_2^{(D)} = r_2^{(UL)}$ must hold for the optimal solution of problem Eq. (4.6).

Meanwhile, the cellular rate functions for uplink sharing can be summarized as

$$
r_m^{(C)} = \begin{cases} \log\left(1 + P_k \left|v_1^H h_{ke}\right|^2\right), & \text{Pure Cellular} \\ 0, & \text{D-B-D} \\ 0, & \text{Dedicated D2D} \\ \log\left(1 + \frac{P_k |v_4^H h_{ke}|^2}{p_4 |v_4^H h_{ie}^{(k)}|^2 + 1}\right), & \text{Underlay D2D} \end{cases} \tag{4.11}
$$

From Eqs. (4.8) and (4.11), we can obtain the following optimal multi-antenna vectors:

$$
\bar{v}_2 = \frac{h_{ie}^{(k)}}{\|h_{ie}^{(k)}\|_2}, \quad \bar{w}_2 = \frac{h_{ei}^{(k)}}{\|h_{ei}^{(k)}\|_2}, \quad \bar{v}_1 = \frac{h_{ke}}{\|h_{ke}\|_2}. \tag{4.12}
$$

For v_4, its optimal form can be derived from

$$
\bar{v}_4 = \arg\max_{v_m} \left\{ \frac{P_k |v_m^H h_{ke}|^2}{p_4 |v_m^H h_{ie}^{(k)}|^2 + 1} \right\}. \tag{4.13}
$$

The expression above is called Rayleigh Quotient [6] and its maximizer is given by

$$
\bar{v}_4 = \frac{\left(p_4 h_{ie}^{(k)} \left(h_{ie}^{(k)}\right)^H + I\right)^{-1} h_{ke}}{\left\|\left(p_4 h_{ie}^{(k)} \left(h_{ie}^{(k)}\right)^H + I\right)^{-1} h_{ke}\right\|_2}. \tag{4.14}
$$

Correspondingly,

$$
r_4^{(C)} = \log\left(1 + P_k h_{ke}^H \left(p_4 h_{ie}^{(k)} \left(h_{ie}^{(k)}\right)^H + I\right)^{-1} h_{ke}\right). \tag{4.15}
$$

Applying eigen decomposition to $p_4 h_{ie}^{(k)} \left(h_{ie}^{(k)}\right)^H + I$, $r_4^{(C)}$ can be further simplified as

$$
r_4^{(C)} = \log\left(b_1 + \frac{b_2}{p_4 b_3 + 1}\right), \tag{4.16}
$$

where $b_1 = 1 + P_k \left(\|h_{ke}\|_2^2 - |\rho|^2\right)$, $b_2 = P_k |\rho|^2$ with $\rho = \left(h_{ie}^{(k)}\right)^H h_{ke}/\|h_{ie}^{(k)}\|_2$, and $b_3 = \|h_{ie}^{(k)}\|_2^2$. Substituting the optimal multi-antenna vectors into the rate functions, the D2D rate can be expressed as $r_m^{(D)} = w_m \log(1 + p_m c_m)$, where $w_1 = w_3 = w_4 = 1$, $w_2 = 0.5$, and c_m's are the D2D channel gain coefficients given by

$$c_m = \begin{cases} 0, & m = 1 \\ \|\boldsymbol{h}_{ie}^{(k)}\|_2^2, & m = 2 \\ |h_i^{(k)}|^2, & m = 3 \\ \frac{|h_i^{(k)}|^2}{P_k|h_{ki}|^2+1}, & m = 4 \end{cases} \cdot \tag{4.17}$$

Further, $r_m^{(C)} = 0$ except that $r_1^{(C)} = \log\left(1 + P_k\|\boldsymbol{h}_{ke}\|_2^2\right)$, $r_4^{(C)} = \log\left(b_1 + \frac{b_2}{p_4 b_3 + 1}\right)$. Plugging the simplified rate functions and using substitution of variable $e_m = x_m p_m$, we can rewrite problem Eq. (4.6) into the following form:

$$\underset{\{x_m, e_m\}}{\text{maximize}} \quad \sum_{m=2}^{4} w_m x_m \log\left(1 + \frac{e_m}{x_m}c_m\right) \tag{4.18a}$$

$$\text{subject to} \quad x_1 r_1^{(C)} + x_4 \log\left(b_1 + \frac{b_2}{\frac{e_4}{x_4}b_3 + 1}\right) \geq R_k \tag{4.18b}$$

$$\sum_{m=1}^{4} e_m \leq E_{ik} \tag{4.18c}$$

$$\sum_{m=1}^{4} x_m \leq 1 \tag{4.18d}$$

$$x_m \geq 0, \; e_m \geq 0 \tag{4.18e}$$

Since $r(x_m, e_m) \triangleq x_m \log\left(1 + \frac{e_m}{x_m}c_m\right)$ is the perspective function of $r_e(e_m) \triangleq \log(1 + e_m c_m)$, the objective function Eq. (4.18a) is joint concave in x_m and e_m. However, $r_4^{(C)}$ is jointly convex in x_4 and e_4. Therefore, Eq. (4.18) is not a convex problem. We will approach the problem using successive convex approximation. Before that, we first discuss problem Eq. (4.6) for downlink sharing scenario.

4.3.2 Downlink Sharing

In downlink sharing, the rate functions $r_1^{(C)}, r_4^{(C)}, r_4^{(D)}$ change whereas other rate functions remain the same as in Eq. (4.18). The optimal $\bar{w}_1$ can be derived similarly as Eq. (4.12). Correspondingly,

$$r_1^{(C)} = \log\left(1 + P_e\|\boldsymbol{h}_{ek}\|_2^2\right). \tag{4.19}$$

From Table 4.1, we can see that w_4 affects both cellular rate $r_4^{(C)}$ and D2D rate $r_4^{(D)}$. Therefore, it is difficult to find the optimum w_4 for problem Eq. (4.6). Since the cellular link has higher priority, we use a suboptimal solution by choosing $\bar{w}_4 = \frac{h_{ek}}{\|h_{ek}\|_2}$ as the maximal ratio combing (MRC) vector that maximizes $r_4^{(C)}$. Consequently,

$$r_4^{(C)} = \log\left(1 + \frac{P_e\|h_{ek}\|_2^2}{p_4|h_{ik}|^2 + 1}\right), \tag{4.20}$$

$$r_4^{(D)} = \log\left(1 + \frac{p_4|h_i^{(k)}|^2}{P_e\left|\left(h_{ei}^{(k)}\right)^H h_{ek}\right|^2 + 1}\right). \tag{4.21}$$

Substituting the rate functions into Eq. (4.6), the problem can be rewritten into a similar form as Eq. (4.18). In the following subsections, we will focus our discussion on Eq. (4.18) noting that it applies to both uplink and downlink sharing scenarios.

4.3.3 Successive Convex Approximation

In this subsection, we will discuss the solution to problem Eq. (4.18). To resolve the non-convexity of the problem, we employ successive convex approximation (SCA) transforming Eq. (4.18) into a series of convex approximated problems. Specifically, the left-hand side (LHS) of Eq. (4.18b) is replaced with its first-order Taylor expansion at different points iteratively. Let $\left(x_4^{(t)}, e_4^{(t)}\right)$ denote the point for Taylor expansion at the t-th SCA iteration. Then $r_4^{(C)}$ can be approximated by

$$\hat{r}_4^{(C)} = r_4^{(C)}\left(x_4^{(t)}, e_4^{(t)}\right) + \left[d_x^{(t)} \quad -d_e^{(t)}\right]\begin{bmatrix} x_4 - x_4^{(t)} \\ e_4 - e_4^{(t)} \end{bmatrix} = x_4 d_x^{(t)} - e_4 d_e^{(t)}, \tag{4.22}$$

where $d_x^{(t)}$ and $d_e^{(t)}$ are expressed in (4.24). The second equality follows from $r_4^{(C)}\left(x_4^{(t)}, e_4^{(t)}\right) = x_4^{(t)} d_x^{(t)} - e_4^{(t)} d_e^{(t)}$. Replacing $r_4^{(C)}$ with $\hat{r}_4^{(C)}$ in problem Eq. (4.18), the convex approximated problem is expressed by

$$\underset{\{x_m, e_m\} \in D}{\text{maximize}} \quad \sum_{m=2}^{4} w_m x_m \log\left(1 + \frac{e_m}{x_m} c_m\right) \tag{4.23a}$$

$$\text{subject to} \quad x_1 r_1^{(C)} + x_4 d_x^{(t)} - e_4 d_e^{(t)} \geq R_k \tag{4.23b}$$

$$\sum_{m=1}^{4} e_m \leq E_{ik} \tag{4.23c}$$

where $D \triangleq \left\{ \{x_m, e_m\} : \sum_{m=1}^{4} x_m \leq 1, \ x_m, e_m \geq 0 \right\}$. Equation (4.23) can be solved efficiently by utilizing available convex optimization tools [7] and the optimum is used for Taylor expansion at the next SCA iteration. It can be shown that the three conditions provided in IV-A of Chiang et al. [8] hold for our problem. Therefore, by iteratively using Taylor expansion and solving the corresponding convex approximated problem, the SCA algorithm will converge to a local maximum of Eq. (4.18).

$$
d_x(x_4^{(t)}, e_4^{(t)}) \triangleq \left. \frac{\partial r_4^{(C)}}{\partial x_4} \right|_{(x_4^{(t)}, e_4^{(t)})} = \log \left(b_1 + \frac{b_2}{e_4^{(t)} b_3 / x_4^{(t)} + 1} \right)
$$

$$
+ \frac{b_2 b_3 x_4^{(t)} e_4^{(t)} (e_4^{(t)} b_3 + x_4^{(t)})^{-1}}{\left[b_1 \left(e_4^{(t)} b_3 + x_4^{(t)} \right) + b_2 x_4^{(t)} \right]} \tag{4.24a}
$$

$$
d_e(x_4^{(t)}, e_4^{(t)}) \triangleq -\left. \frac{\partial r_4^{(C)}}{\partial e_4} \right|_{(x_4^{(t)}, e_4^{(t)})} = \frac{b_2 b_3 \left(x_4^{(t)} \right)^2}{\left[b_1 \left(e_4^{(t)} b_3 + x_4^{(t)} \right) + b_2 x_4^{(t)} \right] (e_4^{(t)} b_3 + x_4^{(t)})} \tag{4.24b}
$$

Recall the discussion in Section 4.3, the objective value of problem Eq. (4.18) as a function of E_{ik} will be used for the resource allocation problem Eq. (4.7). From the analysis above, it is difficult to characterize $Q_{ik}(E_{ik})$ due to that problem Eq. (4.18) is non-convex. Denote the objective value obtained by the SCA algorithm as $\hat{Q}_{ik}(E_{ik})$. We will use it to replace $Q_{ik}(E_{ik})$ in problem Eq. (4.7).

4.4 Resource Allocation

In this section, we will optimize E_{ik} and μ_{ik} for problem Eq. (4.7). Considering that $\{\mu_{ik}\}$ take binary values, we have the following proposition:

Proposition 4.1. Denote $\{\bar{\mu}_{ik}, \bar{E}_{ik}\}$ as the optimal solution of problem Eq. (4.7). It holds that

$$
\bar{E}_{ik} = \bar{\mu}_{ik} \bar{E}_{ik} \tag{4.25}
$$

for all $i \in \mathcal{D}$, $k \in \mathcal{R}$.

Proof. We will prove the proposition for each value of μ_{ik}^*.

1. When $\mu_{ik}^* = 1$, it obviously holds that $E_{ik}^* = \mu_{ik}^* E_{ik}^*$.
2. When $\mu_{ik}^* = 0$, $\mu_{ik}^* Q_{ik}(E_{ik}) = 0$. As a result, E_{ik} can be reduced while increasing $E_{ik'}$ for k' with nonzero $\mu_{ik'}$, which will lead to larger objective value in problem Eq. (4.7) due to the monotonicity of $Q_{ik}(E_{ik})$. Therefore, $E_{ik}^* = 0 = \mu_{ik}^* E_{ik}^*$.

Based on Proposition 4.1, we define new variables $\alpha_{ik} = \mu_{ik}E_{ik}$. Further, replacing $Q_{ik}(E_{ik})$ with $\hat{Q}_{ik}(E_{ik})$, problem Eq. (4.7) becomes

$$\underset{\{\mu_{ik},\alpha_{ik}\}\in F}{\text{maximize}} \quad \sum_{i=1}^{I} \beta_i \sum_{k=1}^{K} \mu_{ik}\hat{Q}_{ik}(\alpha_{ik}/\mu_{ik}) \tag{4.26a}$$

$$\text{subject to} \quad \sum_{k=1}^{K} \alpha_{ik} \leq E_i \qquad \forall\, i \in \mathscr{D} \tag{4.26b}$$

where we define $\mu_{ik}\hat{Q}_{ik}(\alpha_{ik}/\mu_{ik})\Big|_{\mu_{ik}=0} = 0$ and

$$F = \left\{ \{\mu_{ik},\alpha_{ik}\} : \sum_{i=1}^{I} \mu_{ik} \leq 1, \; \forall\, k \in \mathscr{R}, \mu_{ik} \in \{0,1\}, \; \alpha_{ik} \geq 0, \; \forall\, i \in \mathscr{D}, k \in \mathscr{R} \right\}.$$

We will study problem Eq. (4.26) by first relaxing the binary constraints of $\{\mu_{ik}\}$. The binary solution is later enforced based on the study of the relaxed problem.

4.4.1 Lagrangian Dual Decomposition

After relaxing the binary constraint of $\{\mu_{ik}\}$, the Lagrangian function of problem Eq. (4.26) is

$$L(\{\mu_{ik},\alpha_{ik}\},\{\lambda_i\})$$

$$= \sum_{i=1}^{I} \beta_i \sum_{k=1}^{K} \mu_{ik}\hat{Q}_{ik}(\alpha_{ik}/\mu_{ik}) + \sum_{i=1}^{I} \lambda_i \left(E_i - \sum_{k=1}^{K} \alpha_{ik} \right)$$

$$= \sum_{k=1}^{K} \sum_{i=1}^{I} \left[\beta_i \mu_{ik}\hat{Q}_{ik}(\alpha_{ik}/\mu_{ik}) - \lambda_i\alpha_{ik} \right] + \sum_{i=1}^{I} \lambda_i E_i. \tag{4.27a}$$

Its partial derivative with regards to α_{ik} equals

$$\frac{\partial L}{\partial \alpha_{ik}} = \hat{Q}'_{ik}(\alpha_{ik}/\mu_{ik}) - \lambda_i. \tag{4.28}$$

From Eq. (4.28), $\bar{\alpha}_{ik}/\bar{\mu}_{ik}$ that minimizes Eq. (4.27a) can be expressed as a function of λ_i, which we denote as $g(\lambda_i)$. As a result, Eq. (4.27a) can be written as

$$L(\{\mu_{ik},\alpha_{ik}\},\{\lambda_i\}) = \sum_{k=1}^{K} \sum_{i=1}^{I} \mu_{ik} \left[\beta_i\hat{Q}_{ik}(g(\lambda_i)) - \lambda_i g(\lambda_i) \right] + \sum_{i=1}^{I} \lambda_i E_i. \tag{4.29}$$

Since $\mu_{ik} \geq 0$ and $\sum_{i=1}^{I} \mu_{ik} \leq 1$, the optimal i for a fixed k is selected as

$$\bar{i}_k = \arg\max_i \left\{ \beta_i \hat{Q}_{ik}(g(\lambda_i)) - \lambda_i g(\lambda_i) \right\} . \tag{4.30}$$

Correspondingly, $\bar{\mu}_{\bar{i}_k k} = 1$ and $\bar{\mu}_{ik} = 0$ for $i \neq \bar{i}_k$. This indicates that RB chunk k will be allocated to the D2D link with the largest quantity in Eq. (4.30). If $\bar{i}_k$ is unique for all $k \in \mathcal{R}$, then the binary constraint of μ_{ik} will be automatically satisfied.

When $\bar{i}_k$ is not unique, $\bar{\mu}_{\bar{i}_k k}$ may possibly take fractional value, leading to a tie [5]. When ties occur, the value of $\{\bar{\mu}_{\bar{i}_k k}\}$ is determined by the power constraint, which, however, has high computational complexity. Furthermore, the computation of the optimal dual variables $\{\lambda_i\}$ is also prohibitive when the number of D2D links is large. In the following, we will propose a practical RB chunk and energy allocation method that can be implemented distributively, exhibiting low computational complexity, and signaling overhead.

4.4.2 Reduced Complexity Algorithm

Different from [5] in which the rate function admits closed form, evaluating the function value $\hat{Q}_{ik}(\cdot)$ in our problem requires SCA iterations. Therefore, one key factor to reduce the computational complexity of problem Eq. (4.26) is to minimize the number of $\hat{Q}_{ik}(\cdot)$ evaluation. Accordingly, we propose the following Fitting Algorithm:

Initially, we assume uniform energy allocation across all RB chunks. Considering that the energy level depends on the number of RB chunks allocated to each D2D link, we decide the RB chunk allocation by inspecting the objective value of problem Eq. (4.26). Specifically, RB chunk k will be allocated to D2D link i if and only if it results in the largest gain to the weighted D2D sum rate. Let S_i denote the set of currently allocated RB chunks to D2D link i and $|S_i|$ its cardinality, the weighted D2D sum rate gain of allocating RB chunk k to D2D link i is

$$\Delta_i = \beta_i \left[\sum_{k' \in S_i \cup \{k\}} \hat{Q}_{ik'} \left(\frac{E_i}{|S_i| + 1} \right) - \sum_{k' \in S_i} \hat{Q}_{ik'} \left(\frac{E_i}{|S_i|} \right) \right] . \tag{4.31}$$

Assuming $\hat{Q}_{ik}(E_{ik}) = \log(1 + c_{ik} E_{ik})$, where c_{ik} is the effective channel gain of D2D link i on RB chunk k, we have

$$\hat{Q}_{ik} \left(\frac{E_i}{n} \right) \approx \log \frac{1}{n} + \hat{Q}_{ik} (E_i) .$$

Consequently,

$$\Delta_i \approx \beta_i \left[\hat{Q}_{ik} (E_i) + \phi (|S_i|) \right] , \tag{4.32}$$

in which

$$\phi(n) = \log\left(\frac{n^n}{(n+1)^{n+1}}\right).$$

Define $\hat{Q}_k = \max_i \hat{Q}_{ik}(E_i)$. During the RB chunk allocation phase, we first sort the RB chunks in descending order of $\hat{Q}_k$, then assign the RB chunks sequentially. For each RB chunk, the D2D link with the largest $\beta_i\left[\hat{Q}_{ik}(E_i) + \phi(|S_i|)\right]$ will be selected as the optimal link. Since each D2D link only needs to evaluate $\hat{Q}_{ik}(E_i)$ for each RB chunk, the overall complexity of RB chunk allocation in terms of the number of $\hat{Q}_{ik}(\cdot)$ evaluation is IK.

After all the RB chunks are allocated, the energy allocation of D2D link i can be optimized by solving the following problem:

$$\underset{\{E_{ik}\}_{k\in S_i}}{\text{maximize}} \quad \sum_{k\in S_i} \hat{Q}_{ik}(E_{ik}) \tag{4.33a}$$

$$\text{subject to} \quad \sum_{k\in S_i} E_{ik} \le E_i \tag{4.33b}$$

$$E_{ik} \ge 0 \qquad\qquad \forall\, k \in S_i \tag{4.33c}$$

Using general optimization tools to solve problem Eq. (4.33) will be computationally consuming since $\hat{Q}_{ik}(\cdot)$ needs to be evaluated frequently. Recall the approximation $\hat{Q}_{ik}(E_{ik}) = \log(1 + c_{ik}E_{ik})$, we can take $c_{ik} = \left(\exp\left[\hat{Q}_{ik}(E_i)\right] - 1\right)/E_i$. Consequently, problem Eq. (4.33) can be solved by classic water-filling algorithm with

$$\bar{E}_{ik} = \left(\frac{1}{\lambda_i} - \frac{1}{c_{ik}}\right)^+, \tag{4.34}$$

where λ_i can be obtained using bisection search such that $\sum_{k\in S_i} E_{ik} = E_i$.

The Fitting Algorithm above is mainly based on the curve fitting assumption $\hat{Q}_{ik}(E_{ik}) = \log(1 + c_{ik}E_{ik})$. The accurate expression of $\hat{Q}_{ik}(E_{ik})$ depends on the parameters in problem Eq. (4.18) and may not follow this form if more than one mode contributes to the D2D link rate. Nevertheless, owing to mutual interference mitigation achieved by optimized RB chunk allocation, the underlay D2D mode is likely to be dominant in the optimal mode allocation. Therefore, the curve fitting is close to accurate function curve. In the next section, we will show that the proposed algorithm performs sufficiently well in practical D2D underlay system. At the same time, its low computational overhead makes it applicable for real-time network deployment.

Table 4.2 Distributed implementation procedures

1. The BS sends $r_1^{(C)}$, b_1, b_2, b_3, and R_k of each RB chunk to D2D links
2. Each D2D link calculates $\{\hat{Q}_{ik}(E_i)\}_{k \in \mathscr{R}}$ and reports the result to the BS
3. The BS decides the RB chunk allocation and sends the result back to each D2D link
4. D2D links independently implement water-filling algorithm to determine the optimal energy level on each allocated RB chunk
5. Each D2D link optimizes its mode allocation based on the energy level obtained in (4). The resource fraction result is reported back to the BS. The BS further informs CUEs the resource fractions of paired D2D links

4.4.3 Distributed Implementation

In this subsection, we will discuss the implementation procedures of the proposed Fitting Algorithm. Since the BS maintains control of D2D links, the algorithm can be implemented distributively with low signaling overhead using regular LTE downlink and uplink channels. We list the procedures in Table 4.2.

During the RB trunk allocation phase, the BS will acquire $\{\hat{Q}_{ik}(E_i)\}_{k \in K}$ from each D2D link, which is determined locally by solving the mode allocation problem Eq. (4.18). While the D2D link channel state information (CSI) $\{c_m\}$ can be estimated at each D2D transmitter locally, the cellular link CSI parameters $r_1^{(C)}$, b_1, b_2, b_3, R_k of each RB chunk can be acquired from the BS on regular LTE downlink channels. The overall signaling overhead for computing and collecting $\hat{Q}_{ik}(E_i)$ equals $6IK$ in terms of the number of transmitted real quantities. The signaling overhead can be further reduced in a more stable cellular communication environment since the BS only needs to signal the CSI information of those CUEs that recently enter the resource pool and of those experiencing significant changes.

After RB chunk allocation, the BS needs to pass the result to D2D links, each of which can then independently implement water-filling algorithm to optimize the energy allocation across the allocated RB chunks. The overall signaling overhead of passing RB chunk allocation result is K, which can be accomplished through regular LTE downlink channels. Given the optimized energy level across each allocated RB chunk, D2D links can further decide the optimal resource fraction and transmission power of each mode. The resource fraction result is then transferred to the BS, who further passes the information to the corresponding CUE, such that they can coordinate in resource multiplexing for mixed-mode operation. The BS serves as a relay for passing information between D2D links and CUEs. The overall signaling overhead is $6K$ since only three resource fractions need to be informed given the sum constraint (4.18d).

Compared with centralized implementation where the BS optimizes all the variables, the distributed implementation requires less signaling overhead by taking advantage of local processing at the D2D links. Furthermore, it also shifts the

computational load of optimizing mode allocation and energy allocation from the
BS to I D2D links, which enables faster optimization of the system parameters.

4.4.4   Simulation Results and Performance Analysis

In this section, we present numerical results to demonstrate the performance
of the proposed joint mode and resource allocation method. We first illustrate
the advantages of mixed-mode D2D communication in improving cellular rate
constrained D2D link rate. The performance of proposed RB chunk and energy
allocation algorithm is further evaluated in comparison with other algorithms.

4.4.4.1   Mode Allocation Results

We first evaluate the results of the mode allocation problem for a given D2D link and
a given RB chunk. We assume the BS is equipped with four antennas. The channel
vectors are generated randomly from complex Gaussian distribution $CN(\mathbf{0}, d_{ij}^{-\alpha/2}\mathbf{I})$,
where d_{ij} is the distance from transmitter i to receiver j and α is the path loss
exponent. In the simulation, we set $\alpha = 3$.

Figure 4.4 shows the optimized D2D rate with different target cellular rate. Due
to cellular rate constraint, D2D rate naturally drops with the increasing of $r_{C,1}$.
The results demonstrate the significance of impact of cellular QoS constraint on
D2D link performance. Figure 4.5 shows the optimized result under two different

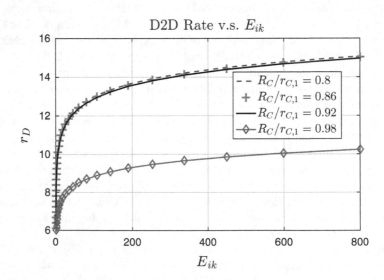

Fig. 4.4 Optimized D2D rate with different target cellular rate

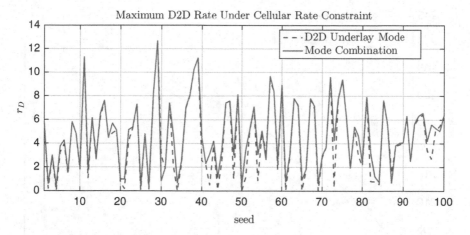

Fig. 4.5 Comparison of mixed-mode transmission with one mode

mode selection strategy, where D2D underlay mode indicates that D2D link can only operate in one mode while mixed-mode allows up to four modes. The curves clearly show the advantages of the mixed-mode transmission strategy in terms of increasing D2D rate under cellular rate constraint.

4.4.4.2   Joint Resource Allocation and Power Control

In Figs. 4.6 and 4.7, we show the simulation results of the reduced complexity algorithm introduced in Sect. 4.4.2. We compared the performance of the algorithm with Greedy Algorithm by exhaustive search. Figure 4.6 shows that rate reduction of our proposed reduced complexity algorithm is less than 2 %, while Fig. 4.7 shows its time reduction is more than 37 %. Therefore, our proposed algorithm is more practical in terms of the performance and complexity tradeoff.

4.5   Chapter Summary

In this chapter, we study the problem of joint mode and resource allocation for mixed-mode D2D enabled cellular networks, where D2D links can multiplex available resources such that they can operate in different modes to meet multiple QoS requirements enforced by the system. We approach the joint optimization problem through two steps. In the first step, the optimal resource fraction and power allocation for different modes are optimized for a fixed pairing between D2D links and cellular resources. In the second step, we study the joint RB chunk and energy allocation problem, which can be casted as a standard resource allocation problem for cellular uplink, whereas the rate functions do not have closed forms.

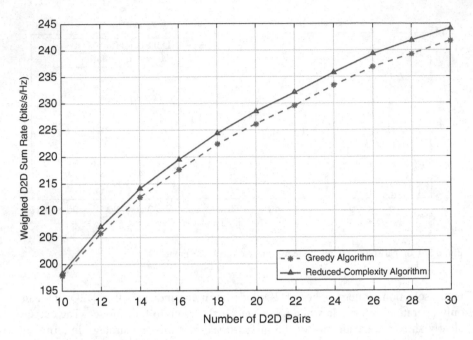

Fig. 4.6 Achievable sum D2D rate

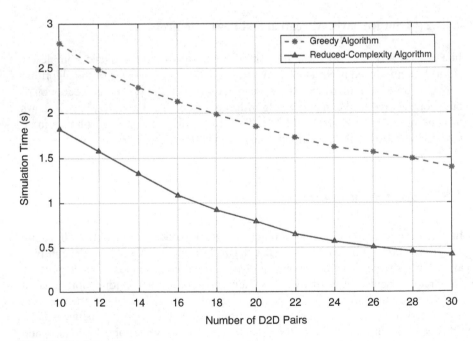

Fig. 4.7 Simulation time

Lagrangian dual decomposition method is employed to solve the problems in both steps, where successive convex approximation is adopted for the non-convex mode allocation problem in the first step and a reduced complexity algorithm is proposed for enabling distributed and practically feasible resource allocation in the second step. Our numerical results demonstrate the practicality of our proposed joint mode and resource allocation scheme. Our study shows that mixed-mode D2D operation can be advantageous in improving heterogeneous D2D network performance. Furthermore, centralized control that is inherent in D2D underlay system enables more efficient and effective network-wide resource deployment.

References

1. K. Doppler, C.H. Yu, C.B. Ribeiro, P. Janis, Mode selection for device-to-device communication underlaying an LTE-advanced network, in *IEEE Wireless Communications and Networking Conference (WCNC)* (2010), pp. 1–6
2. Z. Liu, T. Peng, S. Xiang, W. Wang, Mode selection for device-to-device (D2D) communication under LTE-advanced networks, in *IEEE International Conference on Communications (ICC)* (2012), pp. 5563–5567
3. K. Akkarajitsakul, P. Phunchongharn, E. Hossain, V.K. Bhargava, Mode selection for energy-efficient D2D communications in LTE-advanced networks: a coalitional game approach, in *IEEE International Conference on Communication Systems (ICCS)* (2012), pp. 488–492
4. S. Sesia, I. Toufik, M. Baker, *LTE: The UMTS Long Term Evolution* (Wiley Online Library, 2009)
5. J. Huang, V.G. Subramanian, R. Agrawal, R. Berry, Joint scheduling and resource allocation in uplink OFDM systems for broadband wireless access networks. IEEE J. Sel. Areas Commun. **27**(2), 226–234 (2009)
6. C.B. Chae, I. Hwang, R.W. Heath, V. Tarokh, Interference aware-coordinated beamforming in a multi-cell system. IEEE Trans. Wirel. Commun. **11**(10), 3692–3703 (2012)
7. S.P. Boyd, L. Vandenberghe, *Convex Optimization* (Cambridge University Press, Cambridge, 2004)
8. M. Chiang, C.W. Tan, D.P. Palomar, D. O'Neill, D. Julian, Power control by geometric programming. IEEE Trans. Wirel. Commun. **6**(7), 2640–2651 (2007)

Chapter 5
Resource Management for Cellular D2D Underlay

D2D communications in cellular networks can either work in underlay mode or overlay mode. However, it seems that cellular D2D underlay has attracted more attention than cellular D2D overlay, which may contribute to the high spectrum efficiency brought by spectrum reuse. However, mutual interference caused by spectrum sharing of cellular users and D2D links is an inevitable problem. So it is critical to control the mutual interference and elaborate the scheme of spectrum allocation and power control. In this chapter, resource management in cellular D2D underlay networks is investigated, where D2D links can reuse the spectrum resource of cellular users for data transmission, the so-called resource sharing.

5.1 Critical Problems of Resource Management

In wireless networks, proper resource management in terms of channel allocation and power control is of great significance for interference mitigation and performance improvement. In resource management problems with channel allocation, variables representing channel resource assignment are usually restricted to be integral. For instance, a general definition of this kind of variables can be given as

$$x_{i,j} = \begin{cases} 1, & \text{resource } j \text{ is assigned to user } i. \\ 0, & \text{otherwise.} \end{cases} \tag{5.1}$$

In this case, the optimization problem becomes a combinatorial problem. Generally, there are many kinds of combinatorial optimization problems in wireless resource allocation.

© The Author(s) 2016
L. Wang, H. Tang, *Device-to-Device Communications in Cellular Networks*,
SpringerBriefs in Electrical and Computer Engineering,
DOI 10.1007/978-3-319-30681-0_5

5.1.1 General Problems

For resource allocation in wireless networks, there are many potential applications, and the combinatorial optimization problems can be formulated into different kinds of problems, such as [1]:

Assignment Problems Many optimization problems can be formulated as assignment problems. For instance, one of the typical assignment problems can be described as: a certain set of agents are assigned with a number of tasks, and all the tasks need to be finished by assigning agents in a way that the total cost of the assignment is minimized. The assignment problem can be formulated flexibly with different assumptions. One can suppose that the numbers of agents and tasks are equal and one task can be assigned to exactly one agent to minimize the total cost, or there exist more agents than tasks so that some of the agents may "keep silent doing nothing," or one agent can perform several tasks due to the fact that the number of the tasks exceeds that of the agents. In addition, one can also change the objective of the assignment problem to be profit maximization rather than cost minimization. In cellular D2D underlay, one can describe the spectrum resources of CUEs as agents and D2D links as tasks, and the assignment problem refers to the spectrum matching between cellular resources and D2D links.

Scheduling Problem Scheduling problems are often formulated when specific demands are required at different points in time. Channel assignment problem is one of the representative scheduling problems which requires that specific favorable user is assigned to the channel to maximize the system performance or QoS according to its instantaneous channel conditions or its transmission history.

Knapsack Problem Suppose there are a list of n possible items each with weight w_i and value v_i. One wants to fill a knapsack that can hold a total weight of c with a certain combination of items from the list of n possible items so that the value of the items packed into the knapsack is maximized. This problem is called Knapsack problem, which contains a constraint that the weight of the items in the knapsack do not exceed c, and a restriction that each item can either be in the knapsack or not, i.e., it is impossible to have a fractional amount of the item. When applied in cellular D2D underlay networks, the problem can be described as follows. Suppose that each spectrum resource of CUEs can be shared by several D2D links. However, each CUE has a maximum tolerant interference level to guarantee its own QoS requirement. Each D2D link can reuse the spectrum to improve the system throughput, causing certain interference to the CUE providing the spectrum. Thus, the optimization problem is to maximize the overall system throughput while guaranteeing that the interference caused by D2D links sharing the same resources does not exceed the tolerant interference level.

5.2 Bipartite Graph-Based Resource Management

In this work, we mainly focus on the assignment problem described in Sect. 5.1, to optimize the matching between spectrum resources of CUEs and D2D links in cellular D2D underlay networks. Graph theory is exploited for appropriate resource allocation, particularly by leveraging bipartite graphs.

5.2.1 Graph Matching Problems

Many optimization problems can be represented by graphs defined by nodes and by arcs connecting the nodes. For example, the aforementioned assignment problem can be represented as a bipartite graph by using one set of nodes to denote the agents and using another set of nodes to denote the tasks, and there exists an arc connecting a agent and a task if the agent is capable of finishing that task. Also, in wireless radio resource allocation problems two separate sets of nodes can be used to represent users requiring for communication and spectrum resources, respectively.

Bipartite graphs are leveraged in this part to solve the resource allocation problems, since bipartite graphs are generalized and they can be constructed adaptively according to specific demands. Generally, a bipartite graph can be constructed either weighted or unweighted. In a weighted bipartite graph, the weight of the arcs connecting two nodes from different sets can be defined uniquely according to practical requirements [2]. For instance, a bipartite graph can be constructed to represent the spectrum allocation for DUEs in cellular D2D underlay networks as shown in Fig. 5.1. The spectrum resource of one CUE can be assigned to a number of clustered DUEs for data transmission. Note that when the number of the members in cluster M drops to 1, i.e., $K_M = 1$, the resource allocation problem degenerates to a simple case where the spectrum of each CUE is assigned to one single D2D link. In other words, clustering-based resource matching problem is switched back to a resource pairing problem. To emphasize, the key difference between pairing and clustering problem is that whether it is necessary to consider interference control or spatial reuse for those D2D links within one cluster. As shown in the figure, there exists an edge connecting one spectrum resource and a D2D link/cluster only when certain constraints can be satisfied when allocating that spectrum resource to the corresponding D2D link/cluster. Generally, the constraints can be set as SINR requirements, transmit power limits, transmission success probability, etc. In addition, weights of the edges connecting spectrum resources and DUEs can be defined based on specific objectives, e.g., achievable network capacity, outage probability, energy efficiency, and transmission latency.

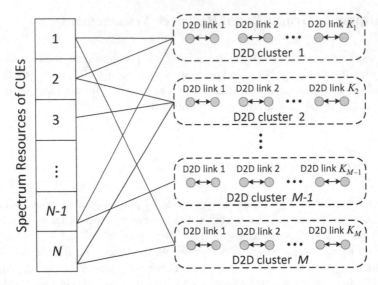

Fig. 5.1 Bipartite graph based resource matching in D2D underlay

5.2.2 Typical Matching Objectives and Solutions

After the construction of bipartite graphs, the matching between nodes from two separate sets can be optimized by using different matching algorithms based on different matching objectives. Taking wireless resource matching between cellular users and D2D links in D2D underlays, for example, three kinds of typical pairing problems are listed as follows, along with the corresponding matching solutions.

5.2.2.1 Maximum Cardinality Pairing Problem

Problem In some cases, one may wish to match as many pairs of nodes as possible. For example, considering the case where the traffic loads are moderate, one may wish to match as many pairs of CUEs (spectrum resources) and D2D links/clusters as possible. In other words, the number of D2D links reusing resources of cellular users is maximized, so as to improve spectrum efficiency in total while guaranteeing basic QoS requirements. By doing so, more D2D links are admitted for communication in cellular D2D underlay networks.

Solution Targeting finding a maximum cardinality matching in a bipartite graph, Hopcroft-Karp (HK) algorithm is one of the most efficient algorithms that can solve the problem described above. Specifically, HK algorithm repeatedly increases the size of a partial matching by finding augmenting paths. However, instead of finding just a single augmenting path per iteration, the algorithm finds a maximal set of shortest augmenting paths. The algorithm terminates when no more augmenting path exists in the bipartite graph. Notice that bipartite graphs are usually unweighted in maximum cardinality pairing problems. The computational complexity of HK is $O(n^{5/2})$, where n is the number of vertices in the bipartite graph [3].

5.2.2.2 Stable Marriage for Maximum Pairing Satisfaction

Problem The well-known stable marriage problem can be described as follows. Given a certain community consists of n men and n women, where each person has ranked all members of the opposite sex in accordance with his or her preferences for a marriage partner, one wants to seek a satisfactory way of marrying off all members of the community such that the marriages are stable. The marriages are defined unstable if there are two people of opposite sex who are not married to each other but prefer each other to their actual mates. When there are no such pairs of people, the set of marriages is deemed stable. The stable marriage problem is particularly applicable in our D2D and CUE pairing. One may wish the resource sharing between CUEs and D2D pairs to be stable in the long term when matching CUEs and D2D links. The stable marriage problem aims to achieve a fair and equitable resource allocation which indicates that each D2D link or CUE should have no incentive to look for better pairings when considering satisfactory performance.

Solution To solve the stable marriage problem, one of the most popular algorithms is Gale-Shapley (GS) algorithm presented by David Gale and Lloyd Shapley in [4]. GS algorithm aims to find a stable matching based on a directed and weighted bipartite graph. Given two sets of nodes, set A and set B with equal number of nodes, the weight of the edge from node a in set A to node b in set B indicates the preference of a to b. In GS algorithm, the nodes in one set (set A) propose to the nodes in another set (set B) based on the preference list. Then each node in set B can hold on to the most favorable proposal it has received. If rejected, the nodes in set A can propose to the next one in the preference list. The algorithm terminates when no more proposals are generated by nodes in set A. In this case, the matching is stable, i.e., no nodes in different sets would rather be matched with each other than remain with the partners that have been assigned to them in the matching. The computational complexity of GS algorithm is $O(n^2)$, where n is the number of nodes in either set.

5.2.2.3 The Optimal Assignment Problem

Problem The optimal assignment problem can be described as a problem of choosing an optimal assignment of n agents to n tasks, assuming that numerical rewards are given for each agent's performance on each task. This problem aims to optimize the match to maximize the sum of the agents' rewards for their assigned tasks. Similarly, the problem can be formulated to minimize the total cost of the agents finishing the tasks, e.g., the total time spend for the tasks. The optimal assignment problem can be transformed into a problem of maximizing/minimizing the sum of the weights of surviving edges on the bipartite graph, which is constructed based on different constraints and objective functions as discussed in Sect. 5.2.1. When applied in cellular D2D underlays, the problem can be described as the maximization of the overall system throughput for all the CUEs and D2D links within the same cell by considering proper pairing decisions.

Solution Kuhn-Munkres (KM) algorithm, originally proposed by H. W. Kuhn in 1955 [5] and refined by J. Munkres in 1957 [6], aims to find the optimal matching in a weighted bipartite graph to optimize the overall system reward. Let A_i and B_j denote nodes from two separated set A and set B, respectively. The algorithm assigns variable α_i to each node A_i and variable β_j to each node B_j. It exploits the fact that the optimization of the assignment problem is feasible when $\alpha_i + \beta_j \leq c_{ij}$, where c_{ij} is the weight of the edge connecting A_i and B_j. An edge in the bipartite graph is called *admissible* when $\alpha_i + \beta_j = c_{ij}$. The subgraph consisting of only the currently admissible edges is called a *equality subgraph*. Starting with an empty matching, the basic strategy employed by the KM algorithm is to repeatedly search for augmenting paths in the equality subgraph. If an augmenting path is found, the current set of matches is augmented by flipping the matched and unmatched edges along this path. Because there is one more unmatched than matched edge, this flipping increases the cardinality of the matching by one, completing a single stage of the algorithm. If an augmenting path is not found, the variables, α_i and β_j, are adjusted to bring additional edges into the equality subgraph by making them admissible, and the search continues. Each stage of the KM algorithm takes $O(n^2)$ arithmetic operations, and the computational complexity of the entire algorithm involving n stages is thus $O(n^3)$, where n is the number of nodes in either set.

5.3   Resource Allocation in Ideal Case

In this section, the optimization of system overall throughput is investigated by leveraging bipartite graph for resource allocation in cellular D2D underlay networks. Furthermore, full CSI of D2D channel is assumed to be known in advance, and the physical channel conditions of D2D links are assumed to be good enough for data transmission. In other words, the D2D links here are always stable enough to afford upcoming transmission. Moreover, the impact of the mobility of mobile users is negligible such that the DUEs can keep in contact during the data transmission.

5.3.1   Model Assumption

Recall that users requesting the same file usually form a cluster so that the desired file can be multicasted within the cluster to save both time and energy resources [7]. Similarly, clusters can be formed by users with common interests for content sharing via D2D communications, since it is more applicable for the situation that users in the same cluster request for the same contents. Generally, the spectrum resource of one CUE can be reused by a single D2D link or several ones in the same cluster in D2D underlay case. However, the interference mitigation problem is more intricate to coordinate among different D2D links in the same cluster. Thus, this chapter is mainly about the resource pairing problem, and our emphasis is about how to reduce complexity without lowering performance noticeably. Therefore, a simple case where at most one D2D link can be allowed to reuse the resource of one CUE at the same time is investigated in the following of this chapter. Moreover, uplink spectrum resources sharing is investigated in this chapter. This is due to the fact that the uplink spectrum is under-utilized in FDD-based cellular system when compared with the downlink. Moreover, only the BS is exposed to the interference caused by DUEs when sharing spectrum on the uplink [8]. Note that, our proposed pairing scheme is also applicable for reusing downlink cellular resources.

In this work, the spectrum sharing for cellular D2D underlay is investigated by focusing on a single cell, subject to their respective minimum QoS requirement. Each active CUE in the cell is assumed to be assigned with one RBG. The main purpose here is to optimize the matching between D2D links and CUEs to share the same RBG. Note that our discussion here assumes the BS will utilize a sophisticated resource allocation scheme to the different CUEs and D2D links, and cochannel CUEs are separated far away geographically so the mutual interference among each other can be negligible.

$\mathscr{C} = \{1, \ldots, N\}$ and $\mathscr{D} = \{1, \ldots, M\}$ denote the index sets of CUEs and D2D links, respectively. In our formulation, the BS is responsible for collecting CSI of relevant CUEs and D2D links in order to optimize the spectrum sharing. Furthermore, the BS can acquire the following information through CUEs and DUEs' reporting or estimation:

- Transmit powers of uplink CUE i and D2D link j, denoted by P_i^c, and P_j^d (limited by their maximum values, $P_{i,\max}^c$ and $P_{j,\max}^d$, respectively);
- Channel gain of D2D link j, channel gain of interference link from CUE i to the receiver of D2D link j, uplink gain between CUE i and the BS, channel gain of interference from D2D link j to the BS, denoted as g_j, $h_{i,j}$, $g_{i,B}$, and $h_{j,B}$, respectively.
- The background AWGN level, denoted as σ^2.

5.3.2   Bipartite Graph Construction

First, a matching indicator variable representing the allocation of cellular resources is defined as,

$$\mu_{ij} = \begin{cases} 1, & \text{if D2D link } j \text{ shares the spectrum resource of CUE } i, \\ 0, & \text{otherwise.} \end{cases} \tag{5.2}$$

In other words, D2D link j has to be assigned dedicated (orthogonal) spectrum for transmission if $\mu_{ij} = 0, \forall i \in \mathscr{C}$. Note that it is assumed that at most one D2D link can share one CUE channel, and one D2D link can share at most one CUE channel at one time. In other words, $\sum_i \mu_{ij} \leq 1$ and $\sum_j \mu_{ij} \leq 1$.

Let f_i^c and g_j^d denote the reward metrics for CUE i and DUE link j, respectively, which can be defined as data rate, spectrum efficiency, energy efficiency, etc. In the following of this chapter, data rate is adopted, for example, to illustrate the resource allocation without loss of generality. Generally, both CUEs and D2D links should have a minimum QoS requirement to guarantee that the mutual interference caused by spectrum sharing would not harm the normal data transmission. Denote the QoS requirements of CUEs and D2D links as $f_i^c \geq f_{i,\min}^c$ and $g_j^d \geq g_{j,\min}^d$. Thus, for each pair of matched CUE i and DUE link j, one must control both P_i^c and P_j^d to maximize the pairwise sum reward,

$$\max_{\mu_{ij}, P_i^c, P_j^d} \quad \sum_{i \in \mathscr{C}} \left(f_i^c + \sum_{j \in \mathscr{D}} \mu_{ij} g_j^d \right), \tag{5.3}$$

$$\text{s.t.} \quad P_i^c \leq P_{i,\max}, P_j^d \leq P_{j,\max},$$

$$f_i^c \geq f_{i,\min}^c, g_j^d \geq g_{j,\min}^d.$$

A bipartite graph can be used for characterizing such pairing problem, as shown in Fig. 5.2. In this graph, the top side involves N cellular users each of which is assigned one resource block group (RBG) for data transmission, while the bottom side involves M D2D users and N virtual partners (one for each CUE). Notice that pairing is not always feasible due to the QoS requirements and power constraints. Some D2D links cannot share resources with any CUEs and dedicated spectrum resources have to be assigned with no resource-sharing reward. For CUEs refusing to share its resource with any D2D links, virtual partners are assigned with no cochannel interference. Hence, there does not need to be an edge linking one CUE resource with one D2D link, which is different from [9, 10].

As shown in Fig. 5.2, the weights of the edges in the bipartite graph can be expressed as

$$w_{i,M+i} = \max_{P_i^c \leq P_{i,\max}^c, P_j^d = 0.} f_i, \tag{5.4}$$

$$w_{i,j} = \begin{cases} \max\limits_{P_i^c \leq P_{i,\max}^c, P_j^d \leq P_{j,\max}^d} (f_i + g_i), & \text{if D2D link } j \text{ can share the spectrum of} \\ & \text{CUE } i \text{ while satisfying the constraints,} \\ 0, & \text{otherwise.} \end{cases}$$

In other words, if CUE i cannot find any D2D links for spectrum sharing while satisfying all the constraints, then CUE i will have only one edge linking to a virtual partner $M + i$ with a weight of $w_{i,M+i}$. Otherwise, there exists an edge linking CUE

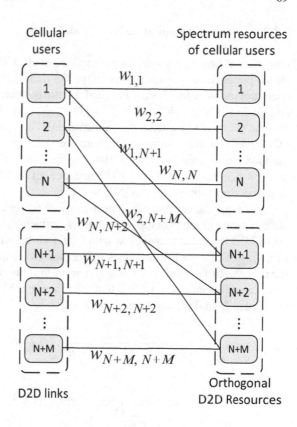

Fig. 5.2 Bipartite graph-based resource pairing

i and each potential D2D link that can satisfy the constraints with a weight of $w_{i,j}$. To emphasize, the main difference between our scheme and others is that we do not force the pairing, since both CUE and D2D links do have their QoS requirements.

5.3.3 Pairing for System Capacity Maximization

To maximize the system capacity by exploiting the aforementioned pairing optimization problem for resource allocation, we first identify a set of candidate D2D links for each CUE for a requisite SINR level to reduce the number of potential pairing searches. Therefore, the computational range will be narrowed to some extent. After that, power control can be executed and pairing algorithms will be used to solve the optimization problem.

5.3.3.1   Determination of D2D Candidate Sets

Our method first establishes an admissible set of D2D links for each CUE, based on the power limits and QoS constraints. For each D2D link, the CUE providing its spectrum resource for sharing is called a reuse partner. Recall that the sets of CUEs and DUEs have been defined as $\mathscr{C} = \{1, \ldots, N\}$ and $\mathscr{D} = \{1, \ldots, M\}$, respectively. We use $\mathscr{D}^i \in \mathscr{D}$ to indicate the candidate D2D set for CUE i, representing all the D2D links that are admittable to share the resource of CUE i while guaranteeing the QoS requirements of both cellular uplinks and D2D links (in terms of SINR). $\mathscr{C}^j \in \mathscr{C}$ denotes the available CUE resource set for D2D j, representing all CUEs that have connecting edges to D2D link j.

Notice that the combinatorial problem with power optimization based on a complete bipartite graph is complex and nearly intractable, our approach is to identify a set of D2D links that can coexist with CUEs while satisfying the QoS requirements of both sides. Some invalid edges in the bipartite graph can be eliminated through the procedure of candidate set determination, thus the computational complexity can be reduced to some extent. Note that the achievable SINR of CUE i and D2D link j, denoted as ξ_i^c and ξ_j^d, can be expressed as

$$\xi_i^c = \frac{P_i^c g_{i,B}}{\sum\limits_{j \in \mathscr{D}} \mu_{ij} P_j^d h_{j,B} + \sigma^2}, \ \forall i \in \mathscr{C} \breve{c} - \forall j \in \mathscr{D}, \tag{5.5a}$$

$$\xi_j^d = \frac{P_j^d g_j}{\sum\limits_{i \in \mathscr{C}} \mu_{ij} P_i^c h_{i,j} + \sigma^2}, \ \forall i \in \mathscr{C} - \forall j \in \mathscr{D}. \tag{5.5b}$$

Based on the calculated SINR, one can determine which of the D2D links can participate in sharing the uplink spectrum resource blocks of each CUE. Thus, our objective is to find the optimal matching result μ_{ij} to maximize the overall system reward. Different from the proposition illustrated in [9], admissible candidate set of D2D links $\mathscr{D}^i$ is selected for each CUE i first. However, the idea for reference of picking candidate CUE set for each D2D link can be borrowed in this work. Since the SINR is reflective of the achievable data rate, here it is taken as the basic mutual constraint to CUE and D2D links to narrow the candidate set. Thus, the candidate set of D2D links for CUE i can be defined as

$$\mathscr{D}^i = \left\{ j \in \mathscr{D} : \xi_j^d(P_i^c, P_j^d) \geq \xi_{j,min}^d \right\}, \tag{5.6}$$

$$\text{s.t.} \ \ \xi_i^c \geq \xi_{i,min}^c, P_j^d \leq P_{j,max}^d, P_i^c \leq P_{i,max}^c, \forall i \in \mathscr{C}, \forall j \in \mathscr{D},$$

where $\xi_{j,min}^d$ and $\xi_{i,min}^c$ denote the minimum required SINR of D2D links and CUEs, respectively. Finding this candidate set requires the BS to check each of the potential D2D links interested in spectrum sharing. Similarly, the active CUE subset for D2D link j is expressed as,

$$\mathscr{C}^j = \{i : i \in \mathscr{C}, \text{ such that } j \in \mathscr{D}^i\}. \tag{5.7}$$

5.3.3.2 Objective of System Capacity Maximization

Recall that f_i^c and g_j^d can be defined to represent data rate, spectrum efficiency, energy efficiency, and so on, in the following of this chapter, we define them to indicate achievable data rate of CUE i and D2D link j, which can be also denoted by R_i^c and R_j^d, respectively. The achievable data rate of CUE i and D2D link j can be expressed as

$$R_i^c = \log_2\left(1 + \xi_i^c\right), \tag{5.8a}$$

$$R_j^d = \log_2\left(1 + \xi_j^d\right). \tag{5.8b}$$

Thus, the optimization problem of maximizing sum rate of all CUE resources is expressed as

$$\max_{\mu_{ij}, P_i^c, P_j^d} \sum_{i \in \mathscr{C}^j, j \in \mathscr{D}^i} \left(R_i^c + \mu_{ij} R_j^d\right) \tag{5.9}$$

5.3.4 Proposed Low-Complexity Pairing Algorithm

In ideal conditions, as illustrated in [10], without taking into the computational cost into account, one optimum solution to the maximum total throughput problem can be achieved through exhaustive pairing search by leveraging the KM algorithm, and power optimization will be executed when evaluating each pairing option on the bipartite graph. However, the algorithm incurs high computational complexity, especially for large number of potential pairings. We now briefly discuss the recently proposed low-complexity inverse popularity pairing order (IPPO) algorithm [11], which is summarized in Algorithm 1.

Aiming to match more spectrum sharing partners of CUEs and D2D links, IPPO starts with the D2D link with the fewest edges (least popular) and finds its best pairing partner with the largest sum rate through joint power optimization for both CUEs and D2D links. We then move down the list with the second D2D link with the next fewest number of edges, and so on. Because the algorithm always goes to the remaining least popular D2D link, it provides better opportunity for overall matching success without iteration (or increasing complexity). This also provides certain fairness assurance to some extent.

Notice that the size of the two partitions of the bipartite graph are not equal, dummy nodes are inserted into the partition with fewer nodes with zero-weight edges connecting all nodes in the opposite partition. Thus, the number of nodes in either partition in Fig. 5.2 becomes $N + M$. As shown in [10], the pairing complexity of KM is $O((N + M)^3)$, whereas that of our proposed IPPO is only $O(N + M)$. More specifically, the achievable overall system rate and the computational time consumed by different algorithms will be illustrated in the next part.

Algorithm 1: Implementation of Proposed IPPO Algorithm

$\mathscr{C}$: the set of CUEs. N: the number of CUEs

$\mathscr{D}$: the set of D2D links. M: the number of D2D links

begin

  Step 1: Set $\mu_{ij} = 0$, $\forall 1 \le i \le N$, $1 \le j \le M$.

  for $i \in \mathscr{C}; j \in \mathscr{D}$

      if both power limitation and SINR requirements are satisfied **then**

          $\mu_{ij} = 1$;

  Step 2: Find number of times each node was selected:

  for $j = 1 : M$ **do**

  $\quad \mid \quad Num_d^j = \sum_{i=1}^{N} \mu_{ij}$,

  end

  for $i = 1 : N$ **do**

  $\quad \mid \quad Num_c^i = \sum_{j=1}^{M} \mu_{ij}$,

  end

  Step 3: Select the D2D link j_0 with minimum Num_d^j:

  for $i = 1 : N$ **do**

  $\quad \mid \quad$ **if** $\mu_{ij_0} = 1$ **then**

  $\quad \mid \quad \quad \mid \quad$ choose CUEi with minimum Num_c^i for D2D linkj_0.

  $\quad \mid \quad \quad \mid \quad$ Optimize powers P_i^c, $P_{j_0}^d$ to maximize $R_i^c + R_{j_0}^d$.

  $\quad \mid \quad$ **end**

  end

  Remove paired nodes from the set. Update N, M.

  Return to Step 3 until $N = 0$ or $M = 0$.

end

Table 5.1 Main simulation parameters

Parameters	Values
Cell radius	300 m
Antenna gain	15 dbi
Noise power (σ^2)	−96 dBm
$P_{i,max}^c$, $P_{j,max}^d$	150 mW, 250 mW
$\xi_{i,min}^c$, $\xi_{j,min}^d$	7 dB, 10 dB
Number of CUEs (N)	10, 15, ..., 30
Number of DUEs (M)	5, 6, ..., 10
Density of PPP	1 nodes/m^2

5.3.5 Simulation Results and Performance Analysis

In our simulation test, the distribution of DUEs follows Poisson point process (PPP). The relevant parameters are shown in Table 5.1 unless specified. We select 100 nodes randomly as the DUEs, from which we pick the closest M pairs of DUEs to form D2D links. Simultaneously, we assume that a variable number of N conventional CUEs are uniformly distributed within the cell.

Fig. 5.3 Sum rate with different number of CUEs and D2D links

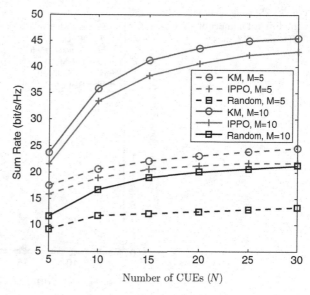

Fig. 5.4 Sum rate with different power limits of CUEs and D2D links

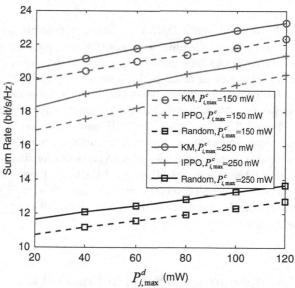

Based on this scenario, we compare IPPO with the well-known KM, as well as the Random pairing algorithm in terms of achieved sum data rate and computation time.

Figures 5.3 and 5.4 illustrate the effects of $P^c_{i,max}$, $P^d_{j,max}$, as well as the number of CUEs and D2D links on the achieved sum rate. Figure 5.3 shows that the sum rate grows when the number of CUEs increases since D2D links would have more choices of potential spectrum resources to share. Similarly, when the number of

Fig. 5.5 Simulation time
with different number of
CUEs and D2D links

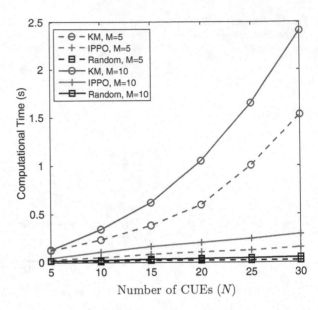

CUEs is fixed, more D2D links make it possible for better pairing with CUE resources, thereby leading to larger sum rate. As shown in Fig. 5.4, larger transmit power limits of CUEs and DUEs lead to higher sum rates. Larger $P^c_{i,max}$ and $P^d_{j,max}$ allow each pair to reach higher rate and hence larger sum data rates. Our IPPO algorithm only exhibits modest performance loss against KM algorithm with high complexity. All of them achieve significantly higher rates than Random pairing.

Figure 5.5 shows that, as CUE number N and D2D link number M grows, the increasing of the computational time consumed by IPPO is negligible when compared to that of the KM, which grows to approximately one order of magnitude higher when N reaches 30. Since LTE scheduling is for each (1-ms) subframes, computation time of several seconds leads to very large scheduling latency. Our results clearly demonstrate the computation advantage of our proposed algorithm without substantial performance loss.

5.4   Resource Allocation in Practical Case

Section 5.3 has illustrated an optimal resource allocation between D2D links and CUEs by considering the circumstance where the physical link condition of D2D pairs is perfect. However, the D2D links are not always stable in practice due to users' mobility and social interactive behavior [12]. In view of that, one critical step is predicting the stability before setting up such links by the served BS. In addition, only statistic channel information for D2D links is assumed to be known. According to this, an admission policy is proposed to determine whether a D2D link can join in the network to transmit data successfully. By measuring the success probability of data transmission, we can obtain the stability level between two DUEs. Accordingly, the criteria can also be used to determine the candidate set of D2D links, and get the optimal spectrum resource allocation.

5.4.1   Social Interaction in Practical Case

In the practical scenario, social interaction among mobile users is also a very important factor when communicating with each other via D2D links. Assume that a pair of DUEs is deployed as a D2D link only if a necessary data block can be delivered successfully. Generally, the information of social interaction can be characterized by using the following two features [13]:

- **Contact Duration:** Due to mobile users' mobility and social behavior, D2D pairs can stay in contact, and actively transmit data file during the time duration. The contact duration measures how long two UEs typically transporting data files when they remain within short distance for D2D connection.
- **Contact Frequency:** The contact frequency measures how often two UEs can meet for D2D data transportation.

From the definition above, one can see that the social interaction model considers not only the time duration of each contact, but also the frequency with which two DUEs can meet, to determine the transmission delay of a data block. Generally, time-sensitive data file must be fully delivered in a single encounter (contact). Packets of lower time-sensitivity, however, can be fragmented and delivered through several encounters. In general, there will be a maximum delay tolerance requirement for fully delivering a data block in the system.

For each of the two aforementioned cases, if a D2D pair can meet the corresponding data delivery requirement, it can be called an admissible pair and the CUE sharing its resource is the spectrum reuse partner. Note that the BS can obtain the useful social interaction information from the operator through historical user activity analysis .

5.4.2   Socially Enabled D2D Link Admission

As we discussed above, social contact time between D2D users is very critical for data transmission via D2D links. In this section, we will focus on D2D link admission problem by considering socially enabled D2D links.

First, delivery success probability $\Pr_s(i, j)$ is defined as the probability that D2D link j can transmit a data block with Z bits by reusing the spectrum resource of CUE

i within the maximum tolerant time. In addition, an admissible D2D link should guarantee that the success probability of data transmission is larger than a given threshold, $v_{\min}^d$.

It is noted that the nominal data block size for D2D delivery can be known proactively by which BS the UE is served, since the packet data network (PDN) gateway can detect the potential D2D traffic, by processing the IP headers of the IP data packets and tunnel headers [14]. Subsequently, two cases are taken into consideration: (1) **one-time encounter** and (2) **multiple encounters** between the D2D partners, and discuss how to select potential D2D pairs to form a candidate set by leveraging both physical link condition and social interaction.

5.4.2.1 Analysis of One-Time Delivery of Time-Sensitive Data Block

Consider a scenario where each pair of DUEs with packets to be transported may not always be in contact (i.e., within the communication range) or connected due to mobility, low power, or user behavior. During the data transmission via D2D links, the contact duration or connection time of two DUEs can be modeled as a continuous random variable, whose distribution can be characterized by its probability density function (PDF). The PDF of the contact duration can be measured by monitoring the social tie history or the closeness of the nodes involved in the D2D link. Here, T is defined as the random variable of the contact duration of D2D links, and its PDF is denoted as $f(T)$.

Recall that the achievable rate has been defined in Eq. (5.8). To satisfy the requirement of successful data delivery, the minimum required contact duration is defined as $\frac{Z}{R_j^d}$. The delivery success probability therefore is defined by

$$\Pr_s(i,j) = \Pr\left\{T \geq \frac{Z}{R_j^d}\right\} = \Pr\left\{R_j^d \geq \frac{Z}{T}\right\}, \tag{5.10}$$

which depends on the achievable rate and the contact duration jointly.

On the other hand, although the BS does not have perfect CSI information of all links in real-time application, it can obtain statistical information of relevant links based on long-term feedback. Specifically, we assume that channel gains, g_j and $h_{i,j}$, follow independent exponential distributions [15], and $u(.)$ denotes the unit step function:

$$f_{g_j}(x) = \frac{1}{\gamma_g} e^{-x/\gamma_g} u(x), \tag{5.11a}$$

$$f_{h_{i,j}}(x) = \frac{1}{\gamma_h} e^{-x/\gamma_h} u(x). \tag{5.11b}$$

Based on the statistical model of channel coefficients and the contact time duration, the success probability expressed in Eq. (5.10) can be derived accordingly.

DUEs must remain in contact for at least a given time threshold to deliver a data block successfully. Similar to [16], exponentially distributed contact time duration T is considered with PDF given by

$$f_T(x) = \frac{1}{\tau} e^{-\frac{x}{\tau}} u(x). \tag{5.12}$$

We first determine the distribution of R_j^d. To specify the group of admissible D2D pairs satisfying the requirement

$$\Pr_s(i,j) = \Pr\left\{R_j^d > \frac{Z}{T}\right\} = \Pr\left\{\frac{P_j^d g_j}{P_i^c h_{i,j} + \sigma^2} \geq 2^{Z/T} - 1\right\} \geq v_{min}^d. \tag{5.13}$$

we first define

$$y \triangleq \frac{P_j^d g_j}{P_i^c h_{i,j} + \sigma^2}, \tag{5.14}$$

its distribution is given in [17] as

Lemma 5.1. *Let x_1 and x_2 follow independent exponential distribution with parameter $\tau = 1$. Then the cumulative distribution function (CDF) of $y = \frac{\alpha x_1}{\beta x_2 + 1}$ with $\alpha, \beta > 0$ is given by*

$$F_y(y) = 1 - \frac{\alpha}{\alpha + \beta y} \exp\left(-\frac{y}{\alpha}\right). \tag{5.15}$$

Let $\alpha = P_j^d \gamma_g / \sigma^2$, $\beta = P_i^c \gamma_h / \sigma^2$, then the CDF of R_j^d can be expressed by

$$F_{R_j^d}(r) = \Pr\left\{R_j^d \leq r\right\} = \Pr\{y \leq 2^r - 1\} = F_y(2^r - 1). \tag{5.16}$$

For now, Eq. (5.13) can be rewritten as

$$\Pr_s(i,j) = \Pr\left\{R_j^d \frac{T}{\tau} \geq \frac{Z}{\tau}\right\} \triangleq \Pr\left\{R_j^d T' \geq \frac{Z}{\tau}\right\}$$

where $T' = \frac{T}{\tau} \sim \text{Exp}(1)$. Using conditional probability, we get

$$\Pr_s(i,j) = \int_0^\infty \Pr\left\{R_j^d T' \geq \frac{Z}{\tau} \,\Big|\, T' = t\right\} \times e^{-t} \, dt$$

$$= \int_0^\infty \left[1 - F_y\left(2^{Z/t} - 1\right)\right] \times e^{-t} \, dt. \tag{5.17}$$

The last equality follows from Eq. (5.16). Substitute Eqs. (5.15) and (5.16) into Eq. (5.17), we have

$$\Pr_s(i,j) \triangleq \Pr_s(\alpha, \beta, Z/\tau) = \int_0^\infty \frac{\alpha}{\alpha + \beta \left(2^{\frac{Z/\tau}{t}} - 1\right)} \exp\left(-\frac{2^{\frac{Z/\tau}{t}} - 1}{\alpha} - t\right) dt.$$

(5.18)

From the expression above, it is easy to verify the following monotonic properties of $\Pr_s(i,j)$:

1. $\Pr_s(i,j)$ is strictly decreasing with β and strictly increasing with α.
2. $\Pr_s(i,j)$ is strictly decreasing with Z and strictly increasing with q.

5.4.2.2 Analysis of Multiple Encounter Delivery with Delay Constraint

In this case, we still consider the same event of transmitting Z bits with a delivery success probability no less than v_{min}^d. However, the transmission can be completed through several encounters (contacts), e.g., resuming download to complete the data delivery. This method does not require continuous contact for transmission, hence allowing more flexible contact duration. However, for practical consideration, the transmission delay cannot be infinitely large. Thus, there is a maximum delay constraint on accomplishing the transmission.

Let δ_{max} indicate the maximum allowable delay for a data block. The number of two user encounters within a given time is modeled as a Poisson process with rate λ. Thus, the distribution of the number of contacts during duration δ_{max} follows:

$$P(K = i) = \frac{e^{-\lambda\delta_{max}}(\lambda\delta_{max})^i}{i!}.$$

(5.19)

Let T_k denote the duration for the k-th contact. Then the success probability for fully delivering Z bits within time duration δ_{max} is given by

$$\Pr_s(i,j) = \sum_{i=1}^\infty \Pr\left\{ \left(\sum_{k=1}^K T_k\right) R_j^d \geq Z \,\Big|\, K = i \right\} \cdot P(K = i).$$

(5.20)

Assume that T_k follows independent exponential distribution given by Eq. (5.12). Then, $T_s \triangleq \sum_{k=1}^K T_k$ follows Erlang distribution, and its PDF is expressed as

$$f_{T_s}(x, K, \tau) = \frac{x^{K-1}e^{-x/\tau}}{\tau^K \Gamma(K)}.$$

(5.21)

Since $\frac{T_s}{\tau} \sim \text{Erlang}(K, 1)$, substituting Eq. (5.21) into Eq. (5.17), the probability for successfully transmitting the data block within K contacts, denoted as $\Pr_s[k]$, is expressed by

$$\mathrm{Pr}_s[k] \triangleq \mathrm{Pr}\left\{\left(\sum_{k=1}^{K} T_k\right) R_j^d \geq Z\right\}$$

$$= \int_0^\infty \frac{\alpha}{\alpha + \beta\left(2^{\frac{Z/\tau}{\tau}} - 1\right)} \exp\left(-\frac{2^{\frac{Z/\tau}{\tau}} - 1}{\alpha}\right) \frac{t^{K-1} e^{-t}}{\Gamma(K)} dt. \qquad (5.22)$$

Further, substituting $\mathrm{Pr}_s[k]$ and the distribution of K into Eq. (5.20), the success probability of multiple encounter delivery can be expressed as

$$\mathrm{Pr}_s(i,j) = \sum_{i=1}^\infty \left\{\left[\int_0^\infty \frac{\alpha}{\alpha + \beta\left(2^{\frac{Z/\tau}{\tau}} - 1\right)} \exp\left(-\frac{2^{\frac{Z/\tau}{\tau}} - 1}{\alpha}\right) \frac{t^{i-1} e^{-t}}{\Gamma(i)} dt\right] \cdot P(K = i)\right\},$$

$$(5.23)$$

To examine the impact of $\delta_{\max}$ on $\mathrm{Pr}_s(i,j)$, we take the derivative of Eq. (5.23) with regards to $\delta_{\max}$, which is expressed by

$$\frac{\partial \mathrm{Pr}_s(i,j)}{\partial \delta_{\max}} = \lambda \left\{\sum_{i=1}^\infty \left[\mathrm{Pr}_s[i] \times \frac{e^{-\lambda\delta_{\max}} (\lambda\delta_{\max})^{i-1}}{(i-1)!}\right] - \sum_{i=1}^\infty \left[\mathrm{Pr}_s[i] \times \frac{e^{-\lambda\delta_{\max}} (\lambda\delta_{\max})^i}{i!}\right]\right\}.$$

$$(5.24)$$

Since $T_k \geq 0$ for all k, it holds that $\left(\sum_{k=1}^i T_k\right) R_j^d \geq \left(\sum_{k=1}^{i-1} T_k\right) R_j^d$. As a result, $\mathrm{Pr}_s[i] \geq \mathrm{Pr}_s[i-1]$. Moreover,

$$\sum_{i=1}^\infty \left[\mathrm{Pr}_s[i] \times \frac{e^{-\lambda\delta_{\max}} (\lambda\delta_{\max})^{i-1}}{(i-1)!}\right] \geq \sum_{i'=1}^\infty \left[\mathrm{Pr}_s[i'] \times \frac{e^{-\lambda\delta_{\max}} (\lambda\delta_{\max})^{i'}}{(i')!}\right] \qquad (5.25)$$

where $i' = i - 1$. Substitute Eq. (5.25) into Eq. (5.24), we get $\partial \mathrm{Pr}_s(i,j)/\partial \delta_{\max} \geq 0$, which indicates that Pr_s is an increasing function of $\delta_{\max}$. Intuitively, looser delay constraint will lead to higher success probability.

5.4.3 Socially Enabled Capacity Maximization Problem

Taking the stability of potential D2D links into consideration, we should reconsider the aforementioned capacity maximization problem. In a practical scenario, the communication stability of D2D links will impact the overall capacity of the system. In other words, the social interaction among mobile users is critical and needs to be taken into consideration. In this case, the candidate set of D2D links can be determined by considering both SINR requirements and the success probability of data transmission within contact time. Therefore, a valid candidate D2D pair reusing CUE i's resource should meet the condition $\Pr_s(i,j) \geq v_{\min}^d$. Thus, the admissible D2D pair set allowed to reuse CUE i's resource can be denoted by

$$\mathscr{D}^i = \left\{ j \in \mathscr{D} : \xi_j^d(P_i^c, P_j^d) \geq \xi_{j,\min}^d, \Pr_s(i,j) \geq v_{\min}^d \right\}, \tag{5.26}$$

$$\text{s.t.} \quad \xi_i^c \geq \xi_{i,\min}^c, P_j^d \leq P_{j,\max}^d, P_i^c \leq P_{i,\max}^c, \forall i \in \mathscr{C}, \forall j \in \mathscr{D}.$$

Similarly, the candidate set of CUEs for D2D link j is denoted by

$$\mathscr{C}^j = \{i : i \in \mathscr{C}, \text{ such that } j \in \mathscr{D}^i\}. \tag{5.27}$$

Thus, based on Eqs. (5.6)–(5.9), our objective function can be expressed as

$$\max_{\mu_{ij}, P_i^c, P_j^d} \sum_{i \in \mathscr{C}^j, j \in \mathscr{D}^i} \left(R_i^c + \mu_{ij} R_j^d \right), \tag{5.28}$$

where $\mathscr{D}^i$ and $\mathscr{C}^j$ are determined in Eqs. (5.26)–(5.27).

Similar to the ideal case, solution of the resource allocation problem in Eq. (5.28) can be obtained by using KM or our proposed algorithm IPPO. Furthermore, $\Pr_s(i,j)$ can be regarded as a parameter which evaluates the trust level among mobile users and can be called *social trust* correspondingly. Such indicator needs to be taken into account when considering the secure communication where the illicit eavesdroppers existed. The detail application of social trust will be illustrated specifically in the following chapter.

5.4.4 Simulation Results and Performance Analysis

Based on our aforementioned analysis of successful delivery for both single-time and multiple encounter D2D links, we now present the achievable system performance based on successful delivery probability. In this simulation part, we set $\lambda\delta_{\max} = 1$, $Z/\tau = 1$, and $v_{\min}^d = 0.5$, and the remaining parameters keep the same as that in Table 5.1 unless specified.

Different from the case illustrated in Sect. 5.3, the information of social interaction plays an crucial role in the stability of D2D links and the success probability of data transmission. In the following simulation results, for notational simplicity, we use "SI" to denote the case where social information is considered to evaluate the stability of D2D links, and "NSI" denotes the case without considering social information.

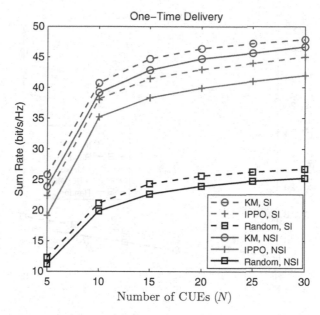

Fig. 5.6 Sum rate with different number of CUEs

5.4.4.1 One-Time Delivery

Figures 5.6 and 5.7 have demonstrated the impact of number of CUEs, power limits of D2D links, and social interaction on the achievable overall sum rate of CUEs and D2D links. In Fig. 5.6, larger N leads to a higher sum rate since D2D links would have more choices of potential spectrum resources to share. With fixed number of user nodes, larger $P^d_{j,\max}$ indicates that each pair has a probability to find a better transmit power to reach higher rate, further leads to larger sum data rate, as shown in Fig. 5.7. Generally, the achievable sum rate of "SI," the case considering the success probability of D2D data transmission exploiting social information, exceeds that of the "NSI." Neglecting the impact of social information, spectrum allocation in "NSI" has a higher probability of pairing DUEs to the CUEs causing lower success probability of data transmission, further leading to difficulty in guaranteeing the achievable sum rate. In addition, similar to the simulation results shown in Sect. 5.3.5, the low-complexity matching algorithm, IPPO, suffers little performance gap when comparing with KM but outperforms the random algorithm to a large extent.

The computational time of varied algorithms with different number of CUEs is illustrated in Fig. 5.8. Intuitively, more simulation time is needed with a higher N for all algorithms, since more edges may survive in the constructed bipartite graph. Apparently, the IPPO is a little more time-consuming than the Random algorithm but costs significantly less time than the KM. Despite the difference of

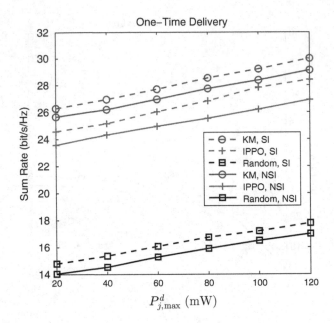

Fig. 5.7 Sum rate with different power limits of D2D links

Fig. 5.8 Simulation time with different number of CUEs

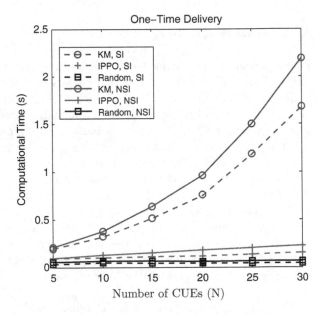

computational time consumed by different algorithms, "SI" is always more time-saving than "NSI," which can be attributed to the removal of the invalid edges with low success probability in the bipartite graph. Combining the results shown in

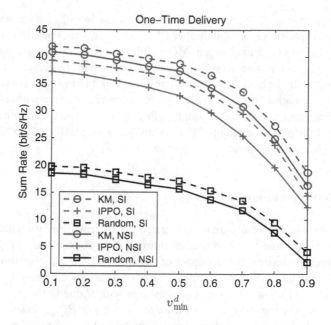

Fig. 5.9 Sum rate with different success probability thresholds

Fig. 5.10 Simulation time with different success probability thresholds

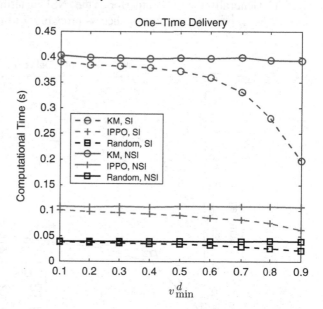

Figs. 5.6, 5.7, and 5.8, the proposed socially based resource management exploiting IPPO strikes a better balance between achievable sum rate and computational time.

Figures 5.9 and 5.10 illustrate the impact of the success probability threshold on the achievable sum rate and computational time, respectively. Apparently, the computational time of "SI" case decreases with increasing $v_{\min}^d$, since fewer edges survive in the bipartite graph with higher $v_{\min}^d$. However, the computational time of "NSI" case almost keeps unchanged. Meanwhile, the achievable sum rate declines with larger $v_{\min}^d$, since narrowing candidate D2D set with larger $v_{\min}^d$ may result in expelling of D2D links with prominent physical link but low success probability. Thus, the determination of $v_{\min}^d$ is critical for complexity reduction without lowering achievable rate noticeably. In addition, the IPPO achieves nearly the same sum rate with KM, and the performance of "SI" case always precedes that of the "NSI" case, which coincides with the results shown above.

5.4.4.2 Multiple Encounter Delivery

In this part, simulation results are demonstrated considering the case where the data transmission is allowed to be completed through several encounters. Figures 5.11 and 5.12 have demonstrated the impact of number of CUEs, power limit of D2D links, and social interaction on the achievable sum rate of CUEs and D2D links. Similarly, Fig. 5.11 shows that sum rate increases with N due to the possibility that better spectrum reusing may exist with larger N. Larger $P_{j,\max}^d$ makes it possible for users to find a better power allocation to achieve better performance as shown in Fig. 5.12. Generally, the "SI" outperforms the "NSI" in terms of achievable sum rate, since "NSI" ignores the impact of success probability of data transmission within

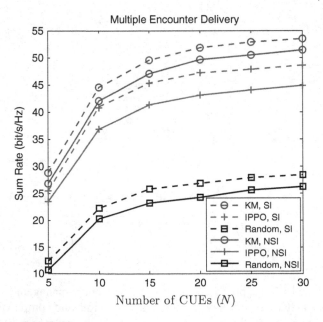

Fig. 5.11 Sum rate with different number of CUEs

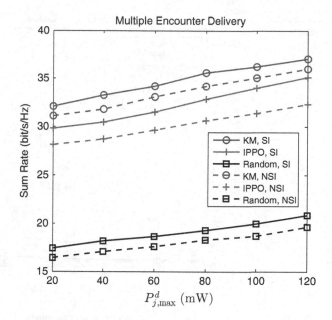

Fig. 5.12 Sum rate with different power limits of D2D links

contact time, leading to a probability of mismanagement of spectrum allocation. Moreover, the computational time depicted in Fig. 5.13 has also elaborated the computational advantage of "SI" owing to the elimination of invalid edges in the constructed bipartite graph. Combining the results shown in Figs. 5.11, 5.12, and 5.13, it can also be concluded that the IPPO acquires nearly the same sum rate but consumes substantially less time when compared with the KM. In contrast, IPPO is superior to the Random concerning sum rate, and the extra time costed by IPPO is negligible.

IPPO algorithm, which strikes a better balance between achievable sum rate and computational time, is leveraged to demonstrate the impact of $\lambda \delta_{\max}$ and $v_{\min}^d$, as shown in Fig. 5.14. Intuitively, the achievable sum rate increases with $\lambda \delta_{\max}$ since larger $\lambda \delta_{\max}$ means that DUEs can meet each other with a higher frequency, or indicates a looser delay constraint. Furthermore, an extremely high threshold of success probability may lead to severe performance loss, because the candidate set determined by a large $v_{\min}^d$ may result in expelling of D2D links with prominent physical link but low success probability.

5.5 Chapter Summary

In this chapter, the uplink spectrum reusing is investigated in cellular D2D underlay networks. To mitigate the mutual interference caused by resource sharing, proper resource management by means of power control and spectrum allocation is

Fig. 5.13 Simulation time
with different number of
CUEs

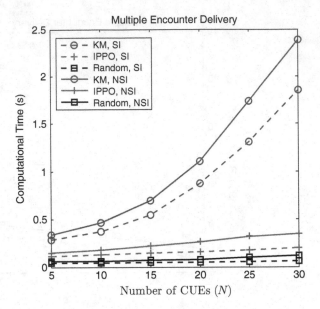

Fig. 5.14 Sum rate with
different success probability
threshold

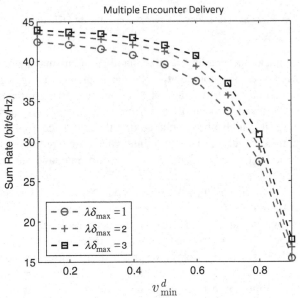

required to improve system performance. Specifically, we solve the optimization
problem of system capacity maximization by exploiting bipartite graph to pair
D2D links with proper cellular resources under certain constraints. In addition
to the ideal case where D2D links are always stable enough to support data
transmission, we also investigate a more practical scenario where D2D links
may experience intermittent outage because of user mobility. To overcome this

uncertainty, we exploit the social interaction of mobile users to make the link more robust. Moreover, an innovative simple pairing algorithm is investigated to strike a balance between achievable data rate and computational time. Numerical results have demonstrated the impact of different parameters on the achievable system performance.

References

1. Z. Han, K.J.R. Liu, *Resource Allocation for Wireless Networks: Basics, Techniques, and Applications* (Cambridge University Press, Cambridge, 2008)
2. L. Wang, H. Wu, W. Wang, K.-C. Chen, Socially enabled wireless networks: resource allocation via bipartite graph matching. IEEE Commun. Mag. **53**(10), 128–135 (2015)
3. J.E. Hopcroft, R.M. Karp, An $n^{5/2}$ algorithm for maximum matchings in bipartite graphs. SIAM J. Comput. **2**(4), 225–231 (1973)
4. D. Gale, L.S. Shapley, College admissions and the stability of marriage. Am. Math. Mon. **69**(1), 9–15 (1962)
5. H.W. Kuhn, The Hungarian method for the assignment problem. Nav. Res. Logist. Q. **2**, 83–97 (1955)
6. J. Munkres, Algorithms for the assignment and transportation problems. J. Soc. Ind. Appl. Math. **5**(1), 32–38 (1957)
7. C. Cao, L. Wang, M. Song, Y. Zhang, Admission policy based clustering scheme for D2D underlay communications, in *2014 IEEE 25th International Symposium on Personal, Indoor and Mobile Radio Communications*, Washington, DC, September 2014, pp. 1937–1942
8. L. Wang, G.L. Stüber, Pairing for resource sharing in cellular device-to-device underlays. IEEE Netw. **30**(2), 122–128 (2016)
9. D. Feng, L. Lu, Y. Wu, G.Y. Li, G. Feng, S. Li, Device-to-device communications underlaying cellular networks. IEEE Trans. Commun. **61**(8), 3541–3551 (2013)
10. J. Han, Q. Cui, C. Yang, X. Tao, Bipartite matching approach to optimal resource allocation in device to device underlaying cellular network. Electron. Lett. **50**(3), 212–214 (2014)
11. L. Wang, H. Wu, Fast pairing of device-to-device link underlay for spectrum sharing with cellular users. IEEE Commun. Lett. **18**(10), 1803–1806 (2014)
12. L. Wang, H, Tang, M. Čierny, Device-to-device link admission policy based on social interaction information. IEEE Trans. Veh. Technol. **64**(9), 4180–4186 (2015)
13. K. Wei, R. Duan, G. Shi, K. Xu, Distribution of inter-contact time: an analysis-based on social relationships. J. Commun. Netw. **15**(5), 504–513 (2013)
14. K. Doppler, M. Rinne, C. Wijting, C.B. Ribeiro, K. Hugl, Device-to-device communication as an underlay to LTE-advanced networks. IEEE Commun. Mag. **47**(12), 42–48 (2009)
15. J. Yue, C. Ma, H. Yu, W. Zhou, Secrecy-based access control for device-to-device communication underlaying cellular networks. IEEE Commun. Lett. **17**(11), 2068–2071 (2013)
16. Y. Zhang, E. Pan, L. Song, W. Saad, Z. Dawy, Z. Han, Social network enhance device-to-device communication underlaying cellular networks, in *Proceedings of 1st International Workshop on Device-to-Device Communications and Networks (D2D 2013)*, Xian, 12–14 August 2013, pp. 182–186
17. H. Min, W. Seo, J. Lee, S. Park, D. Hong, Reliability improvement using receive mode selection in the device-to-device uplink period underlaying cellular networks. IEEE Trans. Wirel. Commun. **10**(2), 413–418 (2011)

Chapter 6
Summary and Future Work

In this book, several practical and critical issues that we may encounter in deploying D2D communications in cellular networks have been studied. Basic introduction and some design aspects and related technologies have been demonstrated in the first two chapters. Then, some application technologies and optimization schemes focusing on proximity discovery, mode selection, and resource management have been proposed and discussed.

The problem of neighbor discovery has been studied for setting up D2D communications in cellular networks. By listening to cellular uplink transmissions, discovering UEs can detect their neighbors and find potential D2D partners in cellular networks. In particular, SRS channel has been utilized as a D2D neighbor discovery opportunity since it is a common uplink channel with potential transmissions from a large number of UEs, among which the discovering UE can identify transmitters in its proximity as candidates for D2D communication. Based on SRS channel structure, we have formulated the problem of neighbor discovery using sparse channel recovery as a unified framework to implement joint neighbor detection and D2D channel estimation. Block sparse Bayesian learning (BSBL) has been employed for maximum likelihood estimation of D2D channel parameters. To improve neighbor detection performance, invariant tests based on composite hypothesis testing have been investigated for designing detectors under a false alarm rate constraint. They can be further combined with BSBL to recover D2D channel statistics.

After the proximity discovery procedure, mode selection is implemented for potential D2D links to determine proper communication mode to enhance system performance. To provide better assistance for mode selection, resource allocation is usually carried out simultaneously. Therefore, the problem of joint mode selection and resource allocation for mixed-mode D2D communications has been studied in cellular networks. Instead of focusing on binary mode selection, this book has assumed that D2D links can multiplex available resources such that they can operate

© The Author(s) 2016
L. Wang, H. Tang, *Device-to-Device Communications in Cellular Networks*,
SpringerBriefs in Electrical and Computer Engineering,
DOI 10.1007/978-3-319-30681-0_6

in different modes to meet multiple QoS requirements enforced by the system. At first, the optimal resource fraction and power allocation for different modes have been proposed for a fixed pairing between D2D links and cellular resources. Then we have studied the joint RB chunk and energy allocation problem, which can be casted as a standard resource allocation problem for cellular uplink, whereas the rate functions do not have closed forms. Lagrangian dual decomposition method has been employed to solve the problems in both steps, where successive convex approximation has been adopted for the non-convex mode selection problem in the first step and a reduced complexity algorithm has been proposed for enabling distributed and practically feasible resource allocation in the second step. The numerical analysis has shown that mixed-mode D2D operation can provide better cellular rate constrained D2D link performance.

Focusing on the case where DUEs work in underlay D2D mode by reusing uplink spectrum resources of CUEs, resource management in terms of spectrum allocation and power control has been studied. We have first listed several general cases of resource management in wireless networks. By exploiting bipartite graphs, different optimization problems and corresponding solutions have been studied in resource management. Specifically, this book targets on the optimization problem of system capacity maximization by exploiting bipartite graph to pair D2D links with proper cellular resources under certain constraints. In addition to the ideal case where D2D links are always stable enough to support data transmission, we have also investigated a more practical scenario where D2D links may experience intermittent outage because of user mobility. To overcome this uncertainty and decide whether two DUEs can work as a pair, social interaction of mobile users has been utilized to evaluate data transmission success probability between users, thereby leading to a more robust resource sharing. Moreover, an innovative simple pairing algorithm has been investigated to strike a balance between achievable data rate and computational time. Numerical results have also demonstrated the effectiveness of our proposed resource management scheme.

Beyond the topics discussed here, we have to say that there still exist many problems in the research field of D2D communications. Our future works may consider additional constraints and provide implementation guidelines for practical D2D operation in cellular networks. These key issues involve: (1) Applying D2D communication into distributed caching networks due to the increasing storage space and computing ability of mobile users; (2) D2D partner selection considering QoS requirements and different levels of delivery success probability; (3) Robust solutions for secure transmission to combat fake cooperative nodes or social outcasts; and (4) Energy efficiency of the cellular inband resource sharing.

In summary, D2D communications can contribute to traffic offloading from BSs, extend network coverage, improve spectrum efficiency, reduce energy consumption, and facilitate many new applications, such as mobile social network, public safety, and moving networks. Beyond this, it will also definitely be exploited in many other areas, not limiting to those aspects we discussed.

Printed in the United States
by BookMasters

Printed in the United States
By Bookmasters